# The Coming Man from Canton

Historical Archaeology of the American West

*Series Editors*

Annalies Corbin
Rebecca Allen

# The Coming Man from Canton

## Chinese Experience in Montana, 1862–1943

Christopher W. Merritt

University of Nebraska Press
*and the*
Society for Historical Archaeology

SOCIETY *for* HISTORICAL ARCHAEOLOGY

Manufactured in the United States of America

∞

Library of Congress Cataloging-in-Publication Data
Names: Merritt, Chris W., author.
Title: The coming man from Canton: Chinese experience in Montana, 1862–1943 / Christopher W. Merritt.
Other titles: Chinese experience in Montana, 1862–1943
Description: Lincoln NE: University of Nebraska Press and the Society for Historical Archaeology, [2017] | Series: Historical archaeology of the American West | Includes bibliographical references and index.
Identifiers: LCCN 2016050487 (print)
LCCN 2016051305 (ebook)
ISBN 9780803299788 (cloth: alk. paper)
ISBN 9781496201201 (epub)
ISBN 9781496201218 (mobi)
ISBN 9781496201225 (pdf)
Subjects: LCSH: Chinese—Montana—History. | Chinese—Montana—Social conditions. | Chinese—Montana—Economic conditions. | Miners—Montana—History. | Railroad construction workers—Montana—History. | Foreign workers, Chinese—Montana—History. | Chinese Americans—Montana—History. | Chinese Americans—Montana—Biography. | Montana—Antiquities. | Montana—Race relations.
Classification: LCC F740.C5 M47 2017 (print) | LCC F740.C5 (ebook) | DDC 978.6/004951—dc23
LC record available at https://lccn.loc.gov/2016050487
Set in Quadraat by John Klopping.

# Contents

# Illustrations

## Tables

## Acknowledgments

Thanks to the dozens of students, faculty, and volunteers who assisted in this complex and sometimes overwhelming project. Without my loved ones this project never would have been so complete, so satisfying, or such a labor of love. Finally, I want to thank the reviewers of this manuscript, the editorial staff, and the copy editor, who spent so many long hours making this the best manuscript possible.

This book is dedicated to those Chinese pioneers who forged a new life in the harsh social and natural climate of Montana and to all those who still come in search of new opportunities under the Big Sky.

# Introduction

*[Though the Chinese] are coolies, and heathen; and "gobble" poor ground and carry away their treasures from our shore, they are yet a laboring element that can be, and is being, used to advantage on this continent. The direction of the labor stream has changed. It crosses now from the Orient, and whether we like or dislike him, Christianize and citizenize him, or he Paganizes us, or neither, "John" is an impending irrepressible fixture on the American Continent and the inevitable Coming Man from Canton.*—New Northwest, *August 13, 1869*

The uncertainty about Chinese immigration expressed in the pages of the *New Northwest* newspaper of Deer Lodge, Montana, encapsulates the ambiguous historical interaction with this population. Montana's vast land and natural resources offered opportunity for all seekers, and the need for cheap labor was a constant in the first decades of the territorial and state history. On the other hand, most Montanans fought over how best to deal legally and socially with the Chinese. Few supported the population's continued habitation in the state. This ambiguity played out daily in Montana's history until the Chinese population dwindled to a level that was no longer seen as a social or economic threat to the state's European American majority.

Since Leeson (1885) published the first comprehensive Montana history, Chinese residents have remained an anecdotal appendix to the state's mainstream narrative. Burlingame's (1942) effort at a statewide historical narrative provided only a cursory examination of the Chinese, focusing on the stereotypical story of their purchasing former European American mining claims that the latter thought had been "panned out." In 1959 K. Ross Toole's thorough discussion of Montana's historical development included only one paragraph about the Chinese population of the state (Toole 1959:72–73). The lack of interest displayed in the Chinese of Montana even by the state's preeminent 20th-century historian reflects the general pattern of scholarly indifference toward this population. Toole's refrain in that one paragraph perpetuated two main stereotypical themes of Chinese history throughout the United States: (1) that they faced an insurmountable hostility from other Montanans, and (2) that they simply worked under-paying placer claims abandoned by non-Chinese miners.

Even with the contribution of Rose Hum Lee's (1949) significant analysis of Chinese immigrant employment succession in Butte between 1890 and 1940, the general flow of Montana historical scholarship bypassed this population. In response to trends set by New Western History (e.g., Rohe 1982; Malone 1989; Malone and Etulain 1989; Malone et al. 1991), the 1980s demonstrated growth in discussions of the Chinese in Montana. Examples of this pattern include John Wunder's (1980) discussion of the legal hurdles faced by Chinese in Montana during the 19th century, Deborah Davis's (1982) analysis of anti-Chinese agitation in the state's newspapers, and Flaherty's (1987) description of the Butte Chinese Boycott. Swartout's (1988) overview of Montana Chinese history from 1862 to 1900 remains a significant contribution to the state's overall understanding of this immigrant group and is the foundation from which this study was built. According to Swartout (1988:53), in contrast to findings in previous studies, the Chinese "played a crucial part in transforming Montana from a primitive, isolated patchwork of localities into an increasingly sophisticated, urbanized, and economically prosperous society."

### The Chinese Migration

In the 19th century alone it is estimated that over 30 million Chinese citizens left their home country in hopes of finding economic prosperity. Internal pressures within China included military and political strife after the Opium Wars in the 1840s, the Taiping Rebellion in the 1860s, famine, and an overall destabilization of centuries-old land-holding traditions (see Pan 1994; Gyory 1998; Kuah-Pearce and Hu-Dehart 2006). Throughout the 19th century most migrants out of China were single or married men who were searching for new economic opportunities. Internal Chinese policy banned the out-migration of women and children, and these rules coupled with foreign policies barring immigration of the same demographics (Kuah-Pearce and Hu-Dehart 2006:4).

Between the 1840s and 1870s millions of Chinese men, and lesser numbers of women and children, ventured to the gold fields of Australia, New Zealand, Canada, and the United States; the copper and silver mines of Chile and Argentina; the sugar plantations of Hawai'i, Jamaica, Trinidad, Guyana, and Cuba; and dozens of other industries in dozens of other countries (McKeown 2001; Pan 1994; Ritchie 1986). Overseas Chinese men would often send or bring back the money earned in these different industries to their families left behind in China. This practice perpetuated a "sojourner" stereotype among host countries and their governments that the Chinese worker was only in the country for material gain and then would take that wealth back to China without much involvement in the local or national economies or politics.

Chinese migrants came from diverse ethnic backgrounds, but most came from the southern and eastern region of China in Guangdong and Fujian provinces (as described more fully in chapter 5). The millions of Chinese immigrants spreading rapidly throughout the world in the 19th century were not bounded by common family lineages, language, cultural tradition, religion, or even social status (Kuah-Pearce and Hu-Dehart 2006). Overseas, the Chinese immigrant population became a single ethnic group due to the external pressures of the host countries and the inherent racism within those societies. At the end of the 19th century many countries passed restrictive immigration policies targeting Chinese immigrants, including the 1882 Chinese Exclusion Act adopted in the United States (Gyory 1998).

While there were small numbers of Chinese citizens living in the United States by the end of the 1700s, the first wave of mass emigration occurred to the newly established gold fields of California in the 1840s–1850s. It was this gold fever, which gripped not only the Chinese but the entire world's population, that led to the common term *Gum San* or Gold Mountain in referring to the United States. By the 1850s rich mineral discoveries radiated outward from California into Nevada, Idaho, Utah, and Montana. With each new discovery the Chinese went in either as miners or as service workers in restaurants, laundries, gardens, or hotels. This migration led to the arrival of the Chinese in Montana in the early 1860s. Montana was but one stop on a globally influential migration of Chinese men and women looking for opportunity. Guotu (2013:40) estimates that there are now roughly 45 million Overseas Chinese spread through 180 countries.

## Archaeological Investigations of Overseas Chinese Communities

Archaeological projects that occurred in several areas of Montana after Swartout's publication added to the growth of historical knowledge regarding the state's Chinese community. Mitigation efforts in the late 1980s resulted in a comprehensive and voluminous discussion of the German Gulch Chinese community, located west of Butte (Fredlund et al. 1991). A decade later Garren Meyer (2001) produced the most significant thesis in the analysis of Montana's Chinese population by using the materials recovered from German Gulch. Additional archaeological projects in the 1990s and early 2000s (Rossillon 1997, 2002, 2003, 2008; Weisz 2003; Bowen 2004; Hall et al. 2006) continued to expand the understanding of the Montana Chinese population, though none attempted to create a more comprehensive statewide context.

While some historical and archaeological parallels existed between Montana's and other diasporic Chinese communities, each area treated this population differently, with some districts showing remarkable acceptance and tolerance (Zhu 1995, 1997). As shown by Zhu (1995:38), the scholarly adherence to a narrative discussing a "dark story of oppression" glossed over significant breaks to this pattern and continually portrayed the Chinese as victims rather than effective actors (Zhu 1995:38).

Archaeological investigations, focused not on stereotypical stories or anecdotes but on material remains of past human activity, provide a less biased and more holistic image of the Overseas Chinese experience in Montana. No one part of the historical or archaeological research is more important than others in understanding the Overseas Chinese experience in Montana. Archaeological data provide a unique and personal glimpse into the daily lives of Montana's Chinese pioneers.

Archaeology also prompts researchers to ask new questions about traditional interpretations of documentary accounts. Newspapers and other historical records place these people and their lives in perspective. These records, while biased in tone, highlight the triumphs, failures, and tribulations of these immigrants from what many in the United States would call a different world. Taken together, archaeological and historical evidence suggest that regardless of their circumstance or persecution, the Overseas Chinese who called Montana home had an active role in shaping the state's specific and broad patterns of history. Montana's Chinese pioneers arrived in Montana during the early part of the 1860s bringing hopes, dreams, and cultural baggage, as did every other wayward immigrant from Ireland, Greece, Spain, England, Sweden, or another of the multitude of nations represented on the pages of the Montana census. What set the Chinese apart was their own unique cultural and social place in the world, sometimes dictated by factors and situations far beyond their control but often based on their own particular struggle to survive and support themselves and the families they left behind.

## Finding the Overseas Chinese Story in Montana

The author's doctoral research began in 2006, aiming for a systematic study of the Chinese in Montana without any existing framework or comparative example of research of this scope within the region. This study is influenced by the concept of historical anthropology as defined by Kent Lightfoot (2005:15, 235): "integration of multiple lines of evidence from documentary, oral, and archaeological sources for a broader, more inclusive view of history." Integrating the sometimes disparate voices of historical and archaeological resources allowed a robust narrative of the Overseas Chinese experience in Montana to emerge from the research. Oral histories represent the most limited pool of research data in this dissertation, as most of the descendant populations no longer live in Montana except for a few families in Helena and Butte.

The initial impetus for the project was a "Chinese terraced garden site" in western Montana, found within an isolated and rugged canyon on the Lolo National Forest. According to local folklore, the dry-laid stonework making up the terraces could only reflect the industry and ingenuity of an Overseas Chinese population fleeing oppression from the nearby Northern Pacific Railroad in the 1880s. After nearly a full year and two excavation projects, the site remains a relative mystery, though historical and archaeological information suggest that it reflects a post-1900 adaptation of a mining site occupied by non-Chinese individuals, likely for moonshining purposes (Merritt 2006, 2009b; Merritt and McLeod 2010).

After this somewhat painful and important lesson in the power of local folklore and stereotypes of rockwork being associated with the Chinese, my dissertation research moved to other archaeological sites associated with this population. Over the next several years, partnerships between the United States Forest Service, Bureau of Land Management, State of Montana, private individuals, and the University of Montana led to research into several dozen other Chinese sites in the state. New investigations included the 1870s mining camps of Louiseville and China Gulch in the Cedar Creek area of Western Montana (Merritt 2007, 2009a), the Big Timber Chinese community (Moschelle 2009), Granite Ghost Town (Blackford and Merritt 2008), a reanalysis of the German Gulch archaeological collection (Norman 2012), recording of a dozen new railroad-related

Chinese encampments along the Clark Fork River (Merritt et al. 2012), and a regional survey and recording effort in partnership with Region 1 of the United States Forest Service (Merritt 2010). During this same period other Montana projects have helped further the understanding of the Chinese, including new historical efforts (Flynn 2006; Stoner 2007; Dean 2009; Robison 2009a, 2009b, 2010) and unaffiliated archaeological projects (Rossillon 2008; McCourt 2010).

## Importance of Comparative Research

This work does not exist in a scholastic vacuum, as dozens of other researchers in the United States, Canada, New Zealand, Australia, and China continue to expand our knowledge and understanding of the Overseas Chinese experience. Much of the historical and archaeological analysis of the Overseas Chinese experience began in the 1970s (e.g., Greenwood 1978; Olsen 1978) and hit the mainstream in the 1980s (Wegars 1993:2). Some pioneering studies into the historicity of the Overseas Chinese population extend back into the early 20th century, but again these saw a fluorescence in the latter part of the 1960s through the 1970s (Lee 1960; Chiu 1967; Sung 1972; Saxton 1975).

Over the relatively short history of Overseas Chinese archaeological studies, the theoretical perspectives and research questions of scholars have shifted substantially. During the initial phases of Chinese research, archaeologists focused on the peculiar, the grand, or the understood (e.g., Teague and Shenk 1977; Olsen 1978, 1983; Hattori et al. 1979; Blanford 1987; Fredlund et al. 1991). Decades after the deconstruction of the culture history concept by proponents of the "New Archaeology" (see Taylor 1948; Binford 1962; Flannery 1967), well into the 1980s many of these initial efforts in Overseas Chinese research still followed the well-worn track of describing artifacts without regard to social or cultural processes. While these early pioneering works furthered the interest of archaeologists attempting to understand these immigrant Chinese populations, the effect of research was largely normative. Greenwood (1993:378) states that most of the archaeological projects until the 1980s focused on reporting of the "whole, glamorous, or recognized" and not on attempting to understand any questions of anthropological importance. There existed no larger framework attempting to understand experiences of Chinese immigrants on a broad scale, and the role of the individual received no attention. One Overseas Chinese research domain impacted by the "New

Archaeology," or Processual movement, was the analysis of acculturation and assimilation (e.g., Moses and Whitmore 1987; Diehl et al. 1998; Xia 2001; Hall et al. 2006). Archaeological discussions of assimilation and acculturation relegated the role of the individual to that of a mere vessel of social norms and cultural values (Voss and Allen 2008:19).

In the last two decades there have been a number of Overseas Chinese archaeological studies that illustrate the shift away from Processual archaeology and toward Post-Processual (see Hodder 1991) analyses that incorporate gender (Voss 2005; Williams 2008), race or ethnicity (Meyer 2001; Voss 2008), funerary rituals and remains (Chung and Wegars 2005; Kraus-Friedberg 2008; Smits 2008), and a broad variety of other research domains (Lydon 1999; Lawrence 2000; Chen 2001; Allen et al. 2002; Bockhorst 2003; Hunt-Jones 2006; Morrow 2009; Ellis et al. 2011). Some archaeologists studying the Overseas Chinese altered their perspectives to incorporate an active agent concept (Praetzellis and Praetzellis 2001; Voss 2008) and looked to expand their knowledge by using ethnographic approaches to get an insider's perspective on the material culture left behind at archaeological sites (e.g., Yang and Hellman 1996; Xia 2001).

Greenwood's (1978) and Wegars's (1993) seminal works provided the research framework for current scholars in the field, and their significant contributions are unquestioned. Work by Ritchie (1986, 1993), Fee (1991, 1993), Chung (1998), Gardner (2000) and Chung and Wegars (2005), illustrate how far research has advanced since the days of Chinese environmental determinism discussed by Spier (1958). There are a number of problems with the continued study of the Overseas Chinese experience. While many historians and anthropologists have moved toward analyses of internal social organization among the Overseas Chinese (e.g., Lyman 1974; Woon 1984; Meyer 2001; Zheng 2001; Liu 2005; Morrow 2009), some archaeologists perpetuate a more simplistic pattern of site-specific histories focused on pitting the Chinese versus the "other," which continues a stereotypical depiction of this immigrant population.

The long-lasting impact of Orientalism was discussed—and critiqued—by Edward Said (1978) and recently reexamined by Madeline Hsu (2003) and Kelly Fong (2005, 2007). Mullins (2008:156) best summarizes the difficulties of Overseas Chinese studies: "Archaeologists face the challenge of constructing an Overseas Chinese subject that respects the authentic cultural roots and experiences of immigrants without devolving into an essentialism based on unreflective analysis of apparently unique material patterns." Indeed, many archaeological examinations have not delved into interpretations dedicated to understanding internal Chinese social organization; rather, archaeological interpretations have focused on the ways in which the Chinese reacted to external stimuli through material culture (assimilation/acculturation). Without a proper understanding of the internal organization of the Overseas Chinese, it is difficult to understand the totality of their experiences. Stanford Lyman (1974), a social historian, studied the social organization of the Overseas Chinese in the 1970s, though it appears many archaeologists forgot those lessons until recently. Voss (2008), Baxter (2008) and Kraus-Friedberg (2008) are some excellent recent examples of how Overseas Chinese archaeological scholarship is transitioning to a true anthropological perspective by focusing on an inside-out versus outside-in perspective on study of this immigrant group. As noted by Voss (2005), only in the last few years has Overseas Chinese research moved toward the mainstream of historical archaeology, due largely to a long-held adherence to assimilation and acculturation models.

### Goals for Studying the Chinese Experience in Montana

Influenced by anthropology, archaeology, and historical threads of scholarly understanding of the Overseas Chinese experience, this study attempts to meet several identified goals ranging from the mundane and foundational to the richly interpretable. First, the lack of a comprehensive statewide historical and archaeological context for the Overseas Chinese has hindered the development of a robust research domain for investigating questions worth asking and continues to do so. The only known successful attempts at creating a macro-scale context for the Overseas Chinese experience have occurred in Australia (Ritchie 2003:7–8) and New Zealand (Ritchie 1986). Working on a regional, or in the case of this study, a statewide level provided both opportunities and limitations in the types of information and analyses possible. Due to the lack of preexisting comprehensive historical analyses of the Montana Chinese population beyond Swartout (1988), the first hurdle was the need to construct a narrative by conducting primary and secondary research for both historical and archaeological materials. The spatial limits in this study created a significant problem of analysis, as using the Montana state boundary inflicts an arbitrary boundary upon a social and natural landscape that in historical reality was not thus bounded. Fortunately, there is a significant body of literature discussing the broader scope of the 19th- and 20th-century Chinese diaspora (Gyory 1998; Ling 1988; Gardner 2000; Dirlick and Yeung 2001; Chan 2006) that allows a more focused spatial scope of analysis, such as the one presented here, to be put into a wider context.

Definition of a temporal boundary was as necessary for this study as defining its spatial limits. The years covered in this book, 1862–1943, are arbitrary but attempt to encapsulate the experiences of the first wave of Chinese immigrants into Montana, beginning with discovery of placer gold along Grasshopper Creek in 1862 and ending with the restructuring of Chinese-American relations during World War II. These are obviously not perfect temporal limits but adequately reflect a shared historical experience by this immigrant population.

A second goal of the study was to approach and interpret the internal organization of the Overseas Chinese population and how this affected their individual and collective lives in Montana. As discussed earlier, only recently have historians and archaeologists attempted to analyze the internal structure of the Overseas Chinese. Instead of focusing on how different this population was from the majority of Montanans, it is important to understand how they structured their lives and relationships to one another within their community when faced with the harsh social and environmental realities of living in the Treasure State. For an archaeologist, material culture—the world of artifacts—constitutes a dynamic interplay between any number of social and cultural mechanisms. In the case of the Overseas Chinese, use of traditional goods could be seen as a refusal to assimilate to mainstream American or Western culture through continued perpetuation of identity markers, but it also reflects an active means within the group to exert control regardless of outside influences. Material culture and historical records tend to bear out the duality of Overseas Chinese experience and in the end provide a richer and more holistic interpretation of their experience in Montana and other parts of the world.

A final goal of this study is to highlight the significant role the Overseas Chinese in Montana played in the development of the state's economy and infrastructure. Swartout (1988) started this process, but over the last 20 years or more the Overseas Chinese story has remained a sidelight to Montana history or a topic to be sprinkled into other stories to add local or exotic color. Partnerships and relationships formed throughout the last few years of undertaking this study have borne fruit in the form of growing awareness of the Chinese Montana experience, which is only furthered by this book. Through funds made available by the University of Montana and Humanities Montana, there are now two websites dedicated to the Montana Chinese experience that will continue to expand as more knowledge and information are gathered.

# The Coming Man from Canton

# 1. Entrance and Expansion, 1862–1880

Weaving together archaeological and historical sources regarding the Chinese experience in Montana completes a more holistic image of the Chinese in the state. Archaeological data, the material remains of Chinese residents, is critical due to the historical prejudices against this population during the 19th and 20th centuries that limited their inclusion in primary-source historical documents. Historical documents from the 1862–1880 period of Montana Chinese history provide some information on where they lived, and for how long, and some anecdotal stories of individual lives. For the most part, what exists in the historical record are generalizations or colorful accounts used to perpetuate racist depictions of the Chinese.

Archaeological studies of sites within this period supply detailed data on the daily lives of these individuals and bring them into better, less biased focus. Together, these two sources are used here to describe the first period of Chinese occupation in Montana, designated as the period of entrance and expansion (1862–1880).

There are currently 20 confirmed archaeological sites recorded in the state with at least some connection to this period of the emergence and establishment of Montana's Chinese population, with an additional 13 sites containing potential for data relevant to this period (Table 1). Unfortunately, many of the sites recorded on this list are heavily disturbed by later episodes of mining or development. Specifically, the archaeological remains of Confederate Gulch and Diamond City are destroyed, though a few miscellaneous Chinese artifacts have been recovered from modern mining claims. Chinese archaeological materials from this period are present at Virginia City, Butte, and Bannack, although due to their relatively urban environments, significant mixing has occurred, placing artifacts out of context. It is not feasible to discuss each recorded site individually; instead, in interpreting early Montana Territory Chinese experience, archaeology and the historical discussions contribute on an equal footing.

This chapter is largely organized as a chronological discussion of the Chinese in Montana, starting with a summary history of Montana before

Table 1. Chinese archaeological sites in Montana relating to the 1862–1880 period

**Confirmed Chinese Archaeological Sites**

| Site # | Site Name | Period | Owner* |
|---|---|---|---|
| 24BE0169 | Bannack | 1862–1880s | State of MT |
| 24BE0187 | Glendale | 1870–1900 | Private/BLM |
| 24BE0807 | Lion City | 1870–1880s | University of MS |
| 24BE1381 | French Creek Cabins | 1870–1880s | USFS |
| 24BW0112 | Confederate Gulch | 1865–1890s | Private/USFS |
| 24JF0059 | Ruddville | 1871–1880 | USFS |
| 24GN0377 | Sierra Mine Site | 1870s–1910s | BLM/Private |
| 24LC0712 | Chinaman's Cove | 1860s–1880s | BOR |
| 24LC1711 | Upper Cave Town | 1868–1880s | USFS |
| 24LC1723 | Upper Cave Gulch | 1870s–1890s | USFS |
| 24MA0723 | Virginia City | 1863–1890s | Private/State |
| 24MN0249 | Louiseville | 1870–1890s | USFS |
| 24MN0262 | China Gulch Hearths | 1870–1875 | USFS |
| 24MN0332 | Rowley Chinese Hearths | 1870–1890s | USFS |
| 24MO0110 | Montreal or "Old Town" | 1874–1890s | USFS |
| 24MO0266 | Fort Missoula | 1877–present | County/USFS/Private |
| 24PW1021 | Blackfoot City | 1865–1880s | Private/BLM/USFS |
| 24SB0212 | German Gulch | 1865–1900 | USFS/Private |
| 24SB0765 | Butte | 1864–1940s | Private |
| 24SB0067 | Highland City | 1865–1870s | USFS |

**Potential or Unconfirmed Chinese Archaeological Sites**

| Site # | Site Name | Period | Owner* |
|---|---|---|---|
| 24BE0806 | Hecla | 1870s–1880s | University of MS |
| 24BE1511 | Cow Creek Placers | 1870s–1880s | USFS |
| 24BE1615 | Trapper City | 1865–1880s | USFS |
| 24BE1872 | Pioneer Cabins | 1870s–1880s | USFS |
| 24BW0287 | Ten Mile Tunnel | 1860s–1880s | BLM/Private |
| 24JF0290 | Chinese Diggings | 1870s–1880s | Private |
| 24JT0075 | Yogo Town | 1878–1880s | USFS |
| 24lc0467 | Old Lincoln | 1865–1880s | USFS |
| 24LC1112 | Austin Mining District | 1870s–1880s | USFS/Private |
| 24LC1716 | Upper Cave Town | 1870s–1880s | USFS |
| 24MO0179 | Yreka | 1860s–1890s | State/UM/Private |
| 24PW0183 | Reynold's City | 1860s–1890s | BLM/Private |
| 24PW0392 | Emmetsburg Cemetery | 1865–1879 | BLM |

* BLM = Bureau of Land Management, USFS = United States Forest Service, BOR = Bureau of Reclamation, UM = University of Montana.

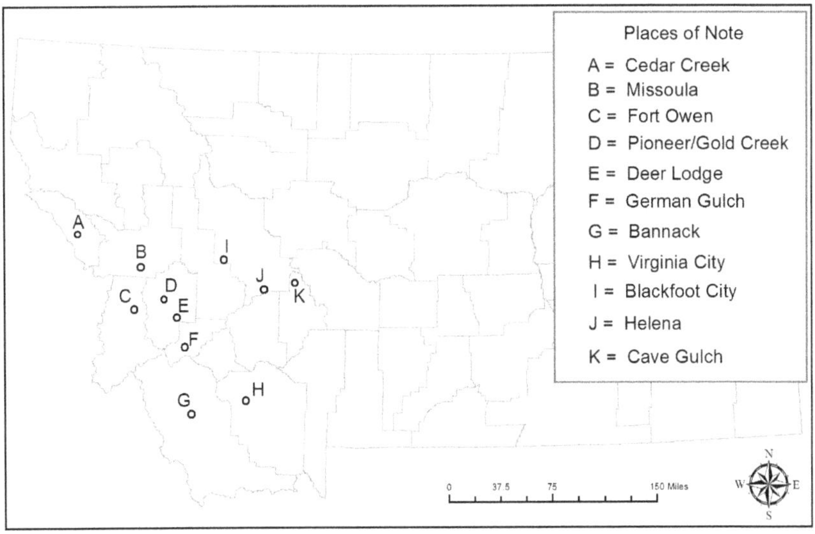


Figure 1: Map of Montana showing location of significant areas in this chapter. (Created by author.)

the gold strikes. After discovery of gold at Bannack in 1862, Chinese immigrants began moving into Montana Territory in substantial numbers (Figure 1 shows locations of areas discussed). By the 1870 federal census, Chinese immigrants accounted for 10% of Montana Territory's population. Between 1870 and 1880 Chinese population centers spread throughout Montana Territory, in many instances overtaking and controlling certain mining districts after European American miners had abandoned the areas. Increasing Chinese population in the territory led, in part, to significant legal challenges being leveled against this large but politically and legally weak population.

With growing racism in Montana Territory's legal system toward the Chinese, a significant facet of Chinese history during this period was how the population handled crime within their own ranks and how crimes against their population were handled in Montana's courts. Finally, the legal precedents and inherent biases in Montana Territory's treatment of the Chinese forced the population to adapt to a significant lack of women in the community. Even with a series of legal challenges during the first 18 years of Chinese history in Montana Territory, courts and judges handled the population on a remarkably level playing field, but this would change by the end of the 1870s. The rapid growth and expansion of their population and the creation of a self-supporting society within Montana Territory exemplify this first period of entrance and expansion (1862–1880).

### Montana Pre-1862

Until the mid-1850s the land that Montana's boundary encompasses today was predominantly a fur trading frontier with sparse populations of trappers, traders, and only a handful of prospectors. The two main settlements were Fort Owen, located in the Bitterroot Valley of western Montana, and Fort Benton at the upper end of the Missouri River in the north-central part of the state. Fort Owen was a trading hub to a small but growing group of Bitterroot Valley farmers and offered some supplies to the travelers of the Emigrant trail heading to California or Oregon (Dunbar and Phillips 1927[1]:2–10). Fort Benton was a center of the fur trade where trappers would bring their harvest for sale and trade (Malone et al. 1991:57).

Before the major gold strikes Montana was decidedly on the fringes of American western expansion. Until 1864 there was no Montana at all, as the lands now constituting the state were split between Idaho and Washington Territory. European American settlements at this time were limited to Fort Owen, Fort Connah, Fort Benton, a loosely organized band of settlers in the Deer Lodge Valley, and a handful of dispersed and relatively insignificant homesteaders (Phillips 1925). As such, the lack of established transportation thoroughfares kept the growth of Montana's settler and prospector populations to a minimum. Survey and construction of the Mullan Road from Fort Benton to Fort Walla Walla in Washington Territory began during the mid- to late 1850s and assisted in developing overland trade in the region (Malone et al. 1991:72). Even with the construction of the Mullan Road, the primary means of moving goods destined for Montana Territory was by steamboat up the Mississippi and Missouri rivers to Fort Benton. From Fort Benton traders took wagon or pack train loads of goods to the various outposts in Montana Territory. Major John Owen was among those involved in early freighting trips. Until settlement of Bannack, and the development of trade routes from Utah and the Emigrant trail, Montana remained an area limited in its ability to prosper by the primitive transportation corridors. The bulk of European American (and later Chinese) interest in Montana did not begin until after gold discovery.

The exact year of the first gold discovery remains unclear, but most agree that between 1852 and 1855 a French fur trapper named Francois

Finlay, commonly called Benetsee by both Native Americans and European Americans at the time, discovered gold in the west-central part of the state near modern-day Gold Creek (Burlingame 1942:79). Finlay had neither the equipment nor ambition to undertake substantial or systematic prospecting. Rather, the gold potential of Montana was not taken seriously until after the arrival of brothers James and Granville Stuart in 1858 (Phillips 1925:133–140). After four years of rumors and growing gold prospecting parties, on July 28, 1862, John White and John McGavin, fresh from the Idaho gold mines, discovered a large pay streak at the mouth of Grasshopper Creek, roughly 20 miles south of modern-day Dillon, Montana (Wolle 1963:50). Within days a rush of miners from across the West began the long trek to Grasshopper Creek from Idaho, Nevada, and California. By the fall of 1862 these miners established the town of Bannack, approximately five miles from the discovery spot, and some estimates suggest the town's population was nearly 2,000 (Wolle 1963:52). By 1863 Major Owen estimated 2,000 or 3,000 additional emigrants entered the area around Bannack largely from Pike's Peak area of Colorado and from Salt Lake (Dunbar and Phillips 1927[1]:281).

The discovery of major placer gold deposits along Grasshopper Creek signaled an end to the early years of Montana's history as primarily a farming and fur trapping frontier. The establishment of Bannack as Montana's first large urban center ushered in this new historical period with pick and pistol. Bannack's colorful early years are not the subject of this chapter in Chinese history, but this was the nature of the social and cultural environment they found when they entered Montana. There is no clear historical evidence that any Chinese were part of the initial flood of immigrants to Bannack in 1862, but it seems a strong likelihood. Regardless of the exact date of their first arrival in the state, the discovery and exploitation of large tracts of placer ground along Grasshopper Creek signaled to the Chinese that Montana had become a land of opportunity.

## Chinese Entrance into Montana 1862–1870

The discovery of gold in California during 1849 prompted the first large-scale migration of Chinese to the United States. By 1850 hundreds of Chinese began arriving on the coast of California to seek their fortunes in the gold fields. By the mid-1850s Chinese had moved into the adjacent gold fields of Oregon and Idaho and into the hard-rock mining communities of Nevada. Within only a decade the Chinese population of the United States had expanded to become tens of thousands strong, with most engaged in mining and mining-related activities along the Pacific coast. It is from these neighboring mining communities in Idaho, Oregon, Nevada, and California that the initial Chinese immigrants came to Montana Territory in the 1860s after the gold strikes.

After discovery of gold in Bannack, prospectors spread throughout southwest Montana in search of the next big strike. In quick succession prospectors found gold along Alder Creek in 1863, and Last Chance Gulch in 1864, and by 1865 miners had founded dozens of placer mining camps throughout the majority of southwest and south-central Montana. Small populations of Chinese began to enter this milieu of entrepreneurial opportunity during the middle 1860s. According to the transcribed minutes of a meeting at the Montana Institute of the Arts History Group in January 1967, Chinese potentially arrived in Virginia City from Bannack in 1863 (Davis 1966:2). According to Davis (1967:2), "When Alder Gulch was discovered I was told that there was one or two Chinamen that left Bannack in 1863 when the gold rush was on and headed for Alder Gulch and I believe this was true." Wiley Davis's information was based on his conversations with his father, who owned a store in Bannack in the early 1860s, so there is potential for error in his recollections; this information must be given a critical eye. The earliest verifiable documentation of Chinese in Montana comes from the pages of the Virginia City newspaper the *Montana Post*. In an article on June 3, 1865, the newspaper editors disparagingly speak of the arrival of a small group of Chinese to Alder Gulch (*Montana Post* 1865:2).

By 1866 descriptions of Chinese began appearing in a variety of historical documents, showing a significant increase in population throughout the territory. The first mention of Chinese in Helena is in the Lewis and Clark County Deed Records, where several groups of Chinese companies began to purchase mining lots in the "Rattlesnake District" in Last Chance Gulch. The deed books record that in 1866 groups of Chinese miners led by Ah Gown, Ah Quay, and Ah Rey purchased several placer claims in upper Last Chance Gulch (Lewis and Clark County Deed Book B:192). In the same year Ah Gih purchased lot 14 in block 3 of the original town plat (Lewis and Clark County Deed Book B:189), which now lies under the public library in downtown Helena. At the time of the Chinese purchase of claims and lots, Helena was part of Edgerton County, named after the first territorial governor of Montana, appointed by President Abraham

Lincoln in 1864. Edgerton County was officially renamed Lewis and Clark County on March 1, 1868. After 1866 there were no more recorded deeds under Chinese names until the 1900s in Lewis and Clark County.

A small community of Chinese moved into the western part of Montana Territory by 1867, as evidenced by entries in Major John Owen's journals. One of the earliest European American settlers in Montana, Major John Owen established Fort Owen in the Bitterroot Valley in the early 1850s. This frontier outpost quickly became a center of commerce and culture for the growing European American population of Montana and also served as Indian Agency for the Salish Tribe into the 1870s. Fort Owen and the farmers surrounding it supplied the initial mining rushes with fresh fruit and vegetables as early as Bannack in 1862 (Spence 1975:130–131).

Major Owen's journals include various mentions of the employment of Chinese cooks to operate the Fort Owen kitchen. It appears Major Owen hired his first Chinese cook, noted as "John Chinaman," from Missoula on June 7, 1867, and paid $110 for his services over a period of 11 weeks (Dunbar and Phillips 1927[2]:58, 68). During his employment at Fort Owen the Chinese cook was apparently accosted by a grist mill employee. One of the cook's eyes was badly swollen, and Major Owen subsequently fired the European American employee (Dunbar and Phillips 1927[2]:67). On August 22, 1867, the Chinese cook hired by Owen returned to Missoula to visit his brother, suggesting at least a small population in the growing agricultural community 35 miles north of Fort Owen (Dunbar and Phillips 1927[2]:68). In October of 1867 Owen hired another Chinese cook, but fired him after only one week; as Owen remarked in his journal, "My Jno China Man Not so good a cook as the one I had during the Summer. He appears Stupid" (Dunbar and Phillips 1927[2]:74–75).

Only a few days after firing the previous cook, Major Owen hired another Chinese cook, though by January of 1868 he mentions two in his journals (Dunbar and Phillips 1927[2]:75, 82). It seems that the same two Chinese cooks appearing in Owen's journals for January 1868 worked through the summer of that year and were rehired the following fall of 1869 (Dunbar and Phillips 1927[2]:82, 144). The last recorded mention of Chinese at Fort Owen is in the 1870 federal census, which lists Jako Lann, a 51-year-old Chinese man, whose occupation is listed as "cooks at [Indian] Agency" (Federal Census Bureau 1870). The presence of Chinese cooks from 1867 to at least 1870 at Fort Owen highlights one type of employment opportunity that Montana Territory offered this immigrant population, and at least tangentially it associates them with the most significant individuals in early state history.

Outside Fort Owen and the Bitterroot Valley in the 1860s, miners discovered large placer deposits in Montana Territory, and with each new rush, the Chinese population increased. By 1868 Raymond (1869:140–141) estimated the Chinese population of Montana Territory to be near 800, with two-thirds of that number employed in placer mining. In 1869 Raymond (1870:260) expanded that estimate to between 2,000 and 3,000 Chinese residents. The significant disparity between Raymond's 1868 and 1869 numbers suggests a less than thorough accounting of the population or a massive increase in the population over the intervening winter.

Regardless of erroneous or conflicting data, in the late 1860s the Chinese population of Montana Territory rapidly expanded, due in large part to the rapid increase in the number of new placer districts. By the late 1860s placer districts such as German Gulch, Gold Creek, Pioneer, Ophir Gulch, Bear Creek, Alder Gulch, Grasshopper Creek, and Last Chance Gulch became large, capitalized, hard-rock mining ventures or were abandoned for new placer strikes elsewhere in the territory. It appears that similar to what happened in other areas of the western United States, Chinese miners were purchasing either abandoned claims or those claims that European Americans could no longer work profitably (Raymond 1869:141; 1870:260). The completion of the Central Pacific Railroad, with its nearest station 290 miles south of Montana Territory's borders, helped to open the region to further settlement by European Americans and Chinese after 1869. Indeed, in the fall of 1869 newspaper reporters noted that in Corinne, Utah Territory, "By the eastern bound train yesterday seventeen more [Chinese] arrived. This makes fifty-eight within four days. They are stopping in Corinne for a few days and will probably proceed on their journey to Montana early the present week" (*Helena Daily Herald* 1869:3).

## Montana Territory's 1870 Chinese Demographics

Montana Territory's first federal census enumeration in 1870 recorded 1,949 Chinese spread across 9 of 13 counties (Figure 2). Major placer mining counties such as Deer Lodge (776), Lewis and Clark (299), and Madison County (122) accounted for over 61% of the entire territory's Chinese population (Table 2). While much has been said about the Chinese population of Montana Territory comprising just 10% of the total (Swartout 1988; Dirlick and Yeung 2001), in the three counties mentioned the pro-

Table 2. Montana Chinese population by county, 1870 Federal Census

| County | Chinese Population | | |
|---|---|---|---|
| | Male | Female | Total |
| Beaverhead | 3 | 3 | 6 |
| Choteau | 3 | 0 | 3 |
| Deer Lodge | 723 | 53 | 776 |
| Gallatin | 2 | 2 | 4 |
| Jefferson | 121 | 1 | 122 |
| Lewis and Clark | 606 | 60 | 666 |
| Madison | 292 | 7 | 299 |
| Meagher | 29 | 0 | 29 |
| Missoula | 42 | 2 | 44 |
| Totals | 1,821 | 128 | 1,949 |

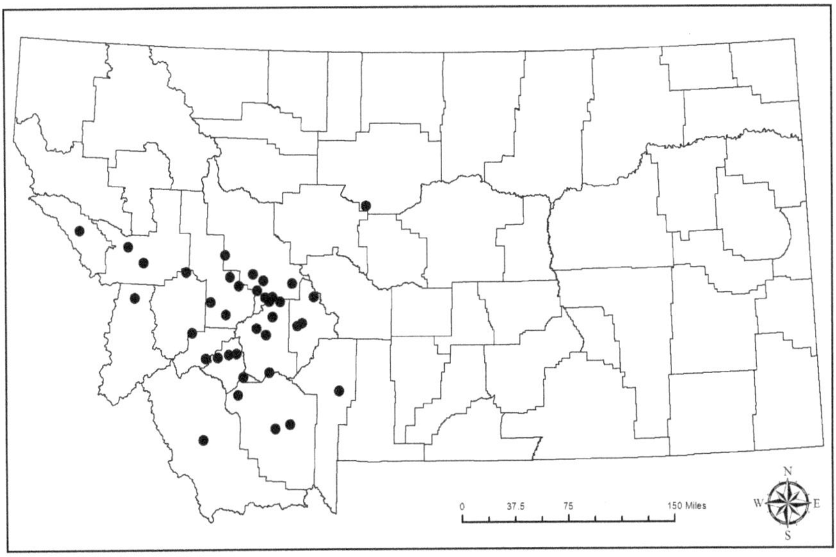


Figure 2: Chinese population centers in 1870. (Created by author.)

portions were higher than the territorial average. In Madison County, which included Virginia City and the surrounding mining camps, Chinese made up 11% of the population. Chinese in Helena and surrounding communities within Lewis and Clark County accounted for 13% of the population. Most significantly, Deer Lodge County had the highest proportion of Chinese in Montana Territory, with their numbers comprising nearly 18% of the population.

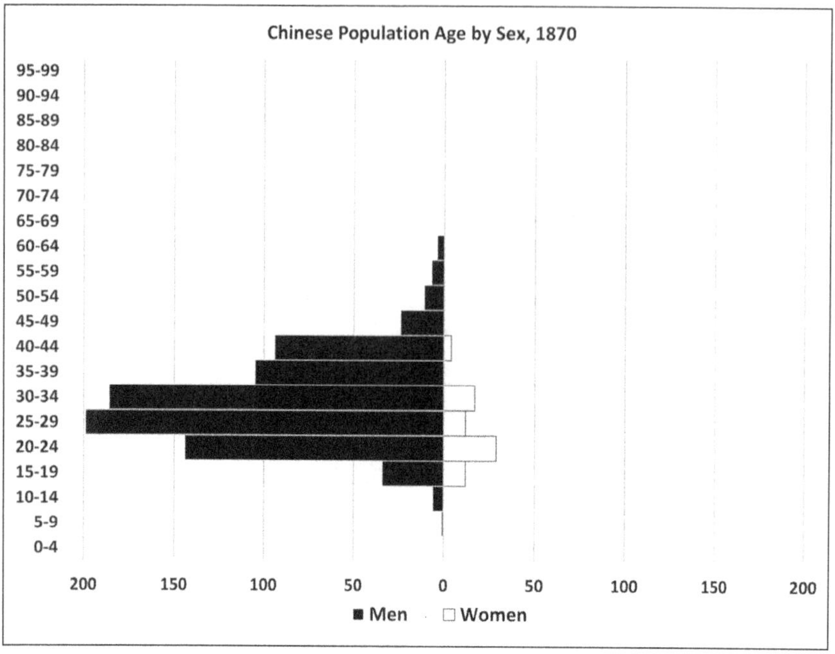


Figure 3: Population pyramid for Chinese in Montana, 1870. (Created by author.)

Of the 1,949 Chinese recorded in 1870, only 128 or 6.6% were women, with the highest numbers in Deer Lodge and Lewis and Clark counties. Interestingly, smaller communities in Beaverhead and Gallatin counties had lower ratios of male to female Chinese but had far smaller total populations. Analysis of individual records within the 1870 census illuminates more patterns in regard to population demographics and employment (Figure 3). The average age of Chinese regardless of sex was 30, although for 1,043 entries, census enumeration had no age data at all. Of the 75 Chinese women with age data, the average was 25 years old, with the median at 24. In contrast, the average age of the 816 Chinese men with associated age data was 30.5 years old, with a median age of 30. The youngest Chinese recorded was a 6-year-old boy in Helena who was apparently born in China, while the youngest female was 16. The oldest Chinese male listed was 60, while the oldest female was 40.

The 1870 census supported the determination by Raymond (1869:140–141) that the majority of Chinese in Montana Territory were engaged in placer mining. In fact, mining accounted for 88.6% of male Chinese employment alone, and 83% of total Chinese employment in the terri-

Table 3. Montana Territory employment of Chinese men, 1870 Federal Census

| Occupation | # of Chinese Employed |
|---|---|
| Mining | 1,613 |
| Laundry Business | 101 |
| Cook | 35 |
| Domestic Servant | 19 |
| Grocer | 9 |
| Store Clerk | 5 |
| Day Laborer | 5 |
| Unknown | 5 |
| Doctor | 4 |
| Baker | 4 |
| Operates Gambling House | 3 |
| Waiter | 3 |
| Saloon Worker | 3 |
| Boarding House Keeper | 3 |
| Boarding House Worker | 2 |
| Tailor | 1 |
| Blacksmith Assistant | 1 |
| Store Keeper | 1 |
| Restaurant Keeper | 1 |
| Gambler | 1 |
| At Home (6-year-old) | 1 |
| "He No Understand" | 1 |
| Unemployed | 1 |

tory. Rohe (1982:13) underestimated the numbers of Chinese placer miners at 70%. Besides the dominant numbers of those working in the mining industry, smaller numbers of Chinese men worked as clothing launderers, domestic servants, and doctors (Table 3). Of the four doctors listed in the 1870 census, three were located in Helena and one was in New York Gulch, northeast of Helena. Dr. Fook On Hong Tong, age 40, provides a clue to Chinese social organization, as it seems more likely that his listed last name, Hong Tong, actually denoted the mutual aid society or fraternal organization he belonged to at the time.

Chinese women had fewer employment options, and for the most part they were "Keeping House" or engaged in prostitution (Table 4). Together these employment spheres accounted for 44.5% and 41.4% of Chinese women's employment respectively. "Keeping House" is a broad term that could suggest either the duties of a homemaker or domestic servant activities at a home for a European American family or a business.

Table 4. Montana Territory employment of Chinese women, 1870 Federal Census

| Occupation | # of Chinese Employed |
|---|---|
| Keeping House | 57 |
| Prostitute | 53 |
| Merchanteel | 7 |
| Laundry Business | 4 |
| Gambler | 2 |
| Cook | 2 |
| Unknown | 2 |
| At Home | 1 |

There is also the potential that a large proportion of the women "Keeping House" were actually prostitutes or concubines, as many census takers used euphemistic terms to identify uncouth employment practices. Further complicating understanding of Chinese women's roles, elsewhere in the mining West, census enumerators in Virginia City, Nevada, mistakenly assumed Chinese women were prostitutes when they were possibly wives or concubines (Chung 1998:220–223). More interesting, perhaps, are the seven women engaged in "Merchanteel" and two noted as gamblers in the 1870 Montana territorial census. It is likely that the seven women, who lived near one another in a section of Virginia City according to the census records, were actually prostitutes in a Chinese brothel or cribs; less likely, they could have been involved in a different type of merchant activity based out of their homes. The two women listed as gamblers, aged 30 and 25, were living with Fook On, a 36-year-old Chinese male operator of a gambling house in Helena. It is unclear what relationship the two women had to Fook On, but again it is possible that they were prostitutes, wives, or had a consanguineal family relationship.

The broad population statistics already noted provide a background on Chinese demographics in 1870 Montana Territory, but macro-scale numbers can gloss over the detail on areas of dense Chinese occupation (Table 5). In 1870 areas like Helena, Virginia City, Pioneer City, and Silver Bow had the bulk of the Chinese population. Helena had the highest proportion of Chinese, with numbers twice those of second-ranked Virginia City. The 1870 census is important as it shows that even early in the Territorial period, Chinese populations clustered in several key locations in and around major placer districts. As noted earlier, Lewis and Clark and Deer Lodge County had significant Chinese populations, yet

Table 5. Montana Chinese population by location, 1870 Federal Census

| City/Town or Locale | # of Chinese Residents |
|---|---|
| Helena | 648 |
| Virginia City | 279 |
| Pioneer City | 245 |
| Silver Bow | 160 |
| Blackfoot City | 111 |
| Butte City | 97 |
| Cable City | 75 |
| Basin | 41 |
| Prickley Pear [sic] | 40 |
| Beartown | 37 |
| Cedar Creek | 34 |
| French Gulch | 27 |
| New York City | 24 |
| Upper Boulder | 19 |
| Rochester | 18 |
| Deer Lodge City | 15 |
| Fish Creek | 12 |
| Trinity/Pidgeon Gulch | 7 |
| Bannack | 6 |
| Greenhorn Gulch | 6 |
| Missoula Village | 6 |
| German Gulch | 5 |
| Springville | 5 |
| Diamond City | 5 |
| Bozeman | 4 |
| Fort Benton | 3 |
| Jefferson City | 3 |
| Unionville | 3 |
| Frenchtown | 3 |
| Lincoln | 2 |
| St. Louis | 2 |
| Madison | 2 |
| Red Mountain City | 1 |
| Washington Creek | 1 |
| Tenmile/Nelson Gulch | 1 |
| Silver Creek | 1 |
| Fort Owen | 1 |

the individual numbers indicate different settlement patterns. For Lewis and Clark County the bulk of the population, 648 of 666 total, lived within the boundaries of Helena itself. More significant than any other statistic discussed already, Chinese accounted for a surprising 20.8% of the total population of the Helena Township. In contrast, in Deer Lodge County, a series of small to moderate Chinese populations dispersed in outlying communities of Pioneer City, Blackfoot City, Cable City, and Beartown, among others.

Perhaps one of the most important Chinese population centers in Deer Lodge County was Deer Lodge itself, though the population was small: only 15. Deer Lodge was important to the Chinese as it appears to have been a nexus of trade for the southwestern part of Montana Territory during the 1870s. As early as fall 1870, Gem Kee & Company opened a store in Deer Lodge with imported goods to cater to both Chinese and European Americans (*New Northwest* 1870d:4). There are other storeowners mentioned in Deer Lodge, including Ah Kane and Quong Lee, but it seems that Gem Kee was a fixture in the town into the 1880s. Monthly freight shipments from Corinne, Utah Territory, indicate the flow of business transacted by Gem Kee, who owned the Kung Chung Lung Company store in Deer Lodge. In May 1877 Kung Chung Lung Company (sometimes written as Kim Chong Lung Company or the Kem Lung Chung store) received a shipment included with 7,136 pounds of goods bound for three customers in Deer Lodge, and the next month the same company was included in a 6,080-pound shipment (*Helena Independent* 1877a:3; 1877b:3). Manifests of steamers coming into Fort Benton on the Missouri River also carried packages for the Kung Chung Lung Company of Deer Lodge in 1878, further illustrating the various ways Chinese goods made it into Montana Territory (*Helena Independent* 1878:3).

It is likely that Deer Lodge, and the various Chinese merchants in that community, served as a central distribution point for goods and facilitated communications between companies across southwest Montana Territory and out-of-territory interests. Indeed, Chinese companies in German Gulch, near Butte, were making the more than 30-mile trek to purchase goods and supplies from Jen Sing, Ah Kane, and Company of Deer Lodge in 1871, much to the chagrin of European American businesses in German Gulch (*New Northwest* 1871g:3). The preference for imported goods from China over domestically available items in host countries is well documented by historians and archaeologists (Greenwood 1978; Ritchie 1986; Fredlund et al. 1991; Wegars 1993; Lawrence 2000; Praetzellis and Praetzellis 2001; Smith 2003). As an indication of the wealth of Chinese merchants in Deer Lodge, Ah Kane returned to China in 1874 for a short visit to his family, which would have been a relatively rare and expensive occurrence in 1870s Montana Territory (*New Northwest* 1874f:3). It appears that Deer Lodge, at least in the early 1870s, was part of a three-distribution-point network in Montana Territory for Chinese goods, while Virginia City and Helena serviced areas around their locations.

## Chinese Population Centers and Mining Districts

During the 1870s there was a remarkable increase in the number of mining districts where Chinese populations dominated, as European American miners moved to higher-paying placer strikes or settled into hard-rock mining ventures elsewhere in the territory. As noted by Rohe (1982:19), the increase of Chinese in a district "did not necessarily correspond with a district's decline. Often it coincided with the exhaustion of the rich surface placers." Even while the economies of numerous countries of the world suffered through a painful slowing of growth during the 1870s, Montana Territory seemed nearly immune. The gold and silver industry of Montana Territory "muffled the impact" of the Panic of 1873, which led to a small depression throughout much of the United States (Spence 1975:39). While the depression of the 1870s slowed the entrance of railroads into Montana Territory, the flow of Chinese continued, but at a slightly decreased rate.

### *Helena*

The Chinese population of Helena was unsurpassed by any other single district in the territory during the 1870s and made up nearly 21% of the residents in that township. It appears that the majority of Chinese in Helena clustered around the downtown area, near the mouth of Last Chance Gulch. Newspaper accounts indicate that Chinese businesses and habitations were in and around the intersections of Cutler, Water, and Main streets. As noted in the 1870 federal census, Helena had a population of 666 Chinese, including 60 women, by far the largest number of Chinese females in the territory. As a result of this significant population, Helena boasted a Chinese-owned and -operated newspaper and a post office by

1870 (*Daily Rocky Mountain Gazette* 1870:2; *Missoula and Cedar Creek Pioneer* 1870a:3). The Chinese newspaper office in Helena was on Cutler Street, and instead of using printing presses, Chinese workers reproduced the volumes using traditional calligraphy by brush and ink (*Daily Rocky Mountain Gazette* 1870:2).

A series of disastrous fires marked the early history of Helena's Chinese community. In April 1869 a fire originating in a Chinese dwelling on the corner of West Main and Bridge Street swept through Helena's business district, destroying dozens of buildings including the offices of the *Helena Herald* (Wolle 1963:74). A few years later and after a second major fire—this time not centered in Chinatown—Helena city officials began to implement a systematic plan to protect the citizens from a major conflagration, including attempts to regulate Chinese housing at the mouth of Last Chance Gulch (*Helena Daily Herald* 1871b:3). The efforts appear to have been ineffective, as on January 17, 1874, a fire again burnt through Helena's business district, destroying at least 150 buildings and with losses estimated at $850,000 (*New Northwest* 1874a:3). The source of that fire was an alleged Chinese gambling house on West Main, and due to severe winds coming out of Last Chance Gulch, ashes were all that was left of most of Chinatown and Helena's downtown. Tong Hing, a Chinese entrepreneur or company, lost three buildings, including the structure where the fire began, totaling a financial loss of $7,000 (*New Northwest* 1874a:3). Since this was the second destructive fire in Helena originating from Chinatown, the European American population's animosity toward the Chinese increased, and editors urged restraint from violence or retaliation in the fire's aftermath (*New Northwest* 1874b:3).

In 1875 the frayed nerves of Helena's European American population were evident in the *Helena Independent*, which ran a story about another fire alarm from Chinatown stemming from fireworks; in this instance a blaze was averted (*Helena Independent* 1875:3). Likely in response to the problems and devastation associated with the earlier fires in Montana and other areas of the West, in 1872 Chinese merchants Ah Kane & Company in Deer Lodge took the initiative and constructed a fireproof building using clay and traditional Chinese tamped-earth building techniques (*New Northwest* 1872d:3). Without question, the series of devastating fires originating in Helena's Chinatown had an unmeasured yet significant effect on policy and planning, forever changing the appearance of the future capital city and other communities.

### *Virginia City*

Virginia City during the 1870s had a Chinese population fluctuating near the 200–300 mark (this was 1/3 of the town's total population), though most of the Chinese residents worked at camps up and down Alder Creek. According to Grant (1998:22) a Virginia City ordinance limited the Chinese to settling in the area of lower Wallace Street, though no other historical support exists for this interpretation. Of the 279 Chinese recorded at Virginia City in 1870, there were seven women, one male cook, and 271 men listed as miners. The Chinese population was significant enough to warrant the construction of a temple and fraternal meeting house as early as the 1860s (Bowen 2004:22). Virginia City's Chinese temple, or "Joss House" as referenced in newspaper articles, was likely the first such building in Montana Territory, though by 1870 there was a temple in Louiseville (*New Northwest* 1870d:4) and probably one in Helena. Chinese miners and merchants located themselves along lower Wallace Street and established a laundry in 1878 (Douglas and Brown 2002:21) and at least two stores by 1884 (Leeson 1885:774). Note that Bowen (2004:22) mistakenly states that Leeson (1885) noted the presence of stores in the late 1870s.

The most commonly told story of Virginia City's Chinese population is of a supposed "Tong War" in the 1870s. It is difficult, even after analyzing historical accounts, to separate fact from fiction when examining this event. Wolle (1963:38–40) provides a detailed account of the battle, yet she provides no citation and no relevant date for the event. The only evidence for a date for the Tong War is indicated by two stories in the *Mineral Independent* of Superior dating to 1938 and 1941, which place the event in 1879. My survey of the *Helena Daily Herald*, *Helena Independent*, and *New Northwest* newspapers for 1879 yielded no corroborating accounts of the supposed Tong War.

Careful analyses of the two stories covered by the *Mineral Independent* highlight major disparities in reporting, including the means of warfare and even the outcome of the conflict. The two news stories share a common thread that the Four Company and Two Company of Chinese miners in Alder Gulch had a conflict over a mining lease that resulted in open warfare and the death of one individual. In the 1938 account the Chinese charged as murderers escaped punishment by the clandestine removal of the body by supporters the night before the trial, thus leading to a mistrial. In the 1941 account of the same event the two Chinese defendants escaped punishment by the help of supporters switching the arrested individuals for others during a prison visit, and when the case came to trial the defense noted that the men in custody were not the ones who had been arrested, and the court dismissed the case. It appears that Wolle (1963:38–40) based her account of the Tong War on the 1938 account, as her discussion repeats—nearly verbatim—the *Mineral Independent* story. This story has now entrenched itself in the psyche of many Montanans, though the original facts have clearly been lost.

### *Cedar Creek*

Only a few months after the completion of Montana Territory's enumeration in the 1870 federal census, an influx of Chinese immigrants altered demographics in the western part of the territory. In the fall of 1869 a pair of French-Canadian prospectors found gold along Cedar Creek, approximately 60 miles west of Missoula (Merritt 2007:5). This was the first major placer strike in western Montana Territory, and thousands rushed to the fledgling district over the winter of 1869–1870. Some estimates place the winter population of 1869–1870 Cedar Creek at nearly 3,000, though the true population was probably much lower (Wolle 1963:270). Newspaper reports exaggerated the mining prospects in the drainage, comparing the potential gold wealth of Cedar Creek to Alder Gulch and Virginia City (*New Northwest* 1869b:3). The 1870 census recorded 34 Chinese individuals in the Cedar Creek area during April of that year, most of whom were either miners or operating stores and businesses at the mouth of Cedar Creek. During the summer of 1870 the population of Chinese immigrants increased by hundreds, due to events in Idaho.

Beginning in the early summer of 1870 European American miners in the Moose Creek Mining District, on the Idaho territorial boundary and directly on the opposite side of the divide from Cedar Creek, passed restrictive rules on the occupation of the district by Chinese (Merritt 2007:9). By 1870 the Moose Creek district was already in decline, which left European American and Chinese miners fighting over the best claim locations (Rice 1977:1–2). European Americans in Idaho passed their own laws within the mining district to the basic effect that no Chinese could hold claims or real property or even live within the district's boundaries. This created an exodus of Chinese from that area, and by October of 1870 at least 300 Chinese men had entered the Cedar Creek drainage from the Idaho Territory, with another 400 expected to immigrate by the end of the year

(*New Northwest* 1870c:3). If the estimates of the population surge in the latter parts of 1870 are near correct, then the total number of Chinese in Montana Territory was between 2,300 and 2,600 by this point, substantially higher than the census of that year would indicate. An editorial in the *Helena Daily Herald* in October 1870 supports this supposition, noting that Chinese "have been coming all summer; have no children or wives here. The females are all prostitutes. There must now be nearly 3,000 Chinese in Montana" (*Helena Daily Herald* 1870c:3).

The initial center of settlement in the Cedar Creek Mining District was Louiseville, approximately halfway between the Clark Fork River and the Idaho territorial border (*Helena Daily Herald*, 1870a). By mid-1870 European American miners moved either higher up the gulch to Forest City and Mayville or to lower camps at Cayuse Gulch, Cedar Junction, and Cedar Mouth, leaving Louiseville mostly abandoned within just a few months of initial construction. Chinese flowing into Cedar Creek from Idaho used the area in and around Louiseville as their initial base of settlement, and by fall 1870 outnumbered European Americans in that community by a two to one ratio. By fall 1870 the Chinese established a store in Louiseville (*New Northwest* 1870c:3). Over the next three years Chinese immigrants established at least one doctor's office, a butcher shop, and a traditional temple in the formerly abandoned town of Louiseville (*New Northwest* 1870c:3, 1870d:4).

Cedar Creek's boom in 1869–1870 led to the first mass migration of Chinese into western Montana Territory and left a lasting physical and archaeological impact on the area. Cultural resource surveys in 1995 and 2008 discovered three sites in the Cedar Creek drainage related to the 1870s–1880s Chinese occupation. Of the original five towns associated with the Cedar Creek gold strike, only one townsite exists today: Louiseville. The only modern evidence of Forest City is a series of hand-stacked rock hearths, depressions, and some ditch works in Mary Anne Gulch, but the main town has long since vanished due to hydraulic mining activities. Due to remining in the 1880s through the 1910s, there are no physical remains of the three other main settlements of Mayville, Cedar Junction, and Cedar Mouth.

An east-west trending tributary of Cedar Creek—currently called China Gulch—is adjacent to downtown Louiseville. In the 1870s China Gulch was called Town Gulch as Louiseville's main street ran from Cedar Creek up this side gulch. In 1995 archaeologists from the Lolo National Forest

Figure 4: Hand-stacked rock hearth (~2 m wide and 1.5 m tall) at China Gulch. (Photo by author, 2008.)

performed archaeological investigations of Louiseville and China Gulch. At Louiseville archaeologists discovered and mapped a dilapidated log cabin dating to the 1870s, a rock building that locals termed the Louiseville jail, and an unassociated 1930s structure that housed a family mining operation through the 1960s. Surface artifacts at Louiseville indicated a mixed deposit of Chinese and non-Chinese archaeological materials. A survey of China Gulch from the eastern boundary of Louiseville located three intact hand-stacked rock hearths (Figure 4) and associated tent or building pads, with an addition of two suspected but unconfirmed hearth features. There were no surface artifacts associated with the features, and cultural affiliation was not determined. Located only a few hundred yards from Louiseville, these hearths likely served as a satellite settlement on the fringes of the larger community. Crews in 1995 excavated two formal units and conducted three smaller shovel tests in Louiseville and an additional four units at one hearth in China Gulch.

Chapter 4 provides discussion of the archaeological findings at China Gulch; to summarize here, there were two significant discoveries from this site. First, excavators located examples of Chinese "funs" trays, created from modified opium cans. Only found at Chinese sites, funs trays are small, uniformly weighted dishes that were used with a scale in the sale of opium. Not identified by the original Forest Service crews, these trays represent the use and sale of opium by residents of China Gulch. A second major discovery, which led to an article published in *Historical Archaeology* (Ellis et al. 2011), was the recovery of several thousand small faunal remains, badly burned and calcined to the consistency of chalk. During analysis Dr. Shannon Novak and Dr. Meredith Ellis of Syracuse University determined that the faunal remains from China Gulch closely match the starvation processed bones of the Donner Party. These material remains of starvation indicate the difficulties faced by the first Chinese to enter this rural part of Montana Territory.

Artifacts recovered at the three excavated hearths cluster around the late 1860s and early 1870s, which fits ideally with the increase in Chinese population after the forced exodus from Idaho. In particular, a Jamaican Ginger bottle dating to the 1860s–1870 and a Drake's Plantation Bitters bottle dating to 1865–1875 suggest a rather brief duration of occupation (Merritt 2007:25, 2009:70). Numerous square (cut) nails and the complete lack of wire (round) nails suggest that the initial Chinese miners scavenged structures in Louiseville to stoke their fires, as there is no physical evidence of any log structures in China Gulch (Merritt 2007:45). A paucity of cultural materials at the hearths suggests that the Chinese living in China Gulch occupied the features only for a few months before moving to better quarters and/or that they were too poor to accumulate a substantial number of goods. Historical documents and archaeological faunal remains support the assertion that the forced exodus from Idaho placed the Chinese migrants in a fragile economic position within Cedar Creek in late 1870, and they were likely only able to improve their status and quarters after working mining claims in Montana.

Archaeological evidence suggests a strong association of the hearths in China Gulch with a short-term Chinese occupation after the migration to Cedar Creek in 1870. Louiseville's archaeological remains, on the other hand, might relate to the second phase of Chinese occupation in Cedar Creek. Excavations in 1995, 2007, and 2008 recovered several thousand artifacts, many of which relate to Chinese occupation. While the historical record indicates that Chinese occupied Louiseville beginning in 1870, the scope and scale of the occupation is unclear. Artifacts recovered from several excavation units imply a long-term occupation, as indicated by a wide variety of Chinese ceramics, opium pipe bowl fragments, opium cans, and "Go" gaming pieces in addition to the non-Chinese bottles and ceramics. Most artifacts recovered from Louiseville suggest a date of 1870–1890s, with some mixing from later occupations in the 1930s–1950s.

When combined, all the archaeological and historical information helps to create a more complete picture of the Chinese experience in Cedar Creek. After their forced migration from Idaho in August 1870, many of the Chinese miners settled in China Gulch, directly adjacent to the European American–dominated and bustling town of Louiseville. After a few harsh weeks or months of scraping by on low food rations, the Chinese moved into permanent quarters by using abandoned cabins and dwellings of Louiseville. Within the next few years Chinese entrepreneurs established a store, a butcher shop, and possibly a doctor's office in the town to cater to the several hundred Chinese miners working claims in the center of the Cedar Creek district. The settlement remained on the periphery of major settlements in Idaho and Montana territories, and Chinese store owners imported only small and easily transported ceramic vessels (and the goods therein) and opium cans. Domestic animals including cows and pigs dominated the diet of the Louiseville Chinese, following a pattern found in other archaeological settings (Wegars 1993; Rossillon 2003). By the end of the 1870s the Chinese population dipped to approximately 130 and continued to decline throughout the next decade. Chinese accounted for 50% of the population in the Cedar Creek Mining District in 1880 (Federal Census Bureau 1880). By 1890 there were only a handful of Chinese working the claims in Cedar Creek, the boom long since having left Louiseville and its last few residents behind.

### Ninemile Creek

In 1874 the same French-Canadian prospectors who had discovered gold in Cedar Creek also hit paydirt in the Ninemile District, 15 miles west of Missoula. In addition to the hundreds of European American miners from Cedar Creek who rushed to the new Ninemile strikes, Chinese also moved to the area to establish laundries and purchase under-performing claims. The initial boom of the Ninemile District in the mid-1870s spurred construction of two new settlements, Martina and Montreal (or "Old Town"), but by 1880 there were no Chinese recorded in the district. During the initial (1874–1875) boom of Ninemile District, Montreal had nearly a dozen European American–owned businesses and two Chinese laundries (Wolle 1963:268). By 1876 newspaper accounts stated that Chinese miners were the "monarch of all they survey" in Eustache Gulch, a tributary of Ninemile Creek just upstream from Montreal (*Helena Independent* 1876c:3).

There is little articulated historical information available on the Chinese on Ninemile Creek, and the overall history of mining district is remarkably poor. Archaeological materials recorded by the Lolo National Forest are the only significant evidence of the Chinese population of Montreal, found at the confluence of St. Louis, Eustache, and Ninemile creeks. In 2000 crews battling a massive wildfire in the Ninemile drainage cut a bulldozer line through the downtown section of the previously recorded Montreal archaeological site. A few days later archaeologists surveyed the bulldozer cut and noted the presence of Chinese artifacts, including fragments of opium cans, celadon vessels, and brown-glazed stoneware consistent with a liquor jar (Roenke 2000:1). Bulldozer crews had accidentally encountered the archaeological remains of the Chinese laundries or associated habitations existing at Montreal in the late 1870s.

### German Gulch

Throughout the early 1870s Chinese began purchasing large tracts of previously worked placer claims in districts across Montana Territory. Nearly every time a new Chinese purchase occurred, a newspaper carried an editorial with the details. Beginning with the Cedar Creek strike discussed earlier, Chinese began purchasing large numbers of claims throughout previously developed areas. In the 1870 census German Gulch, west of Butte, had only five Chinese individuals listed.

The first mention of transactions in German Gulch came in April 1871, when negotiations began between Chinese and European American companies for purchase of placer claims for a total of $17,000 (*New Northwest* 1871c:3). Within the next few months, Chinese companies purchased several claims from the European American–owned Union Mining Company, totaling 1,600 feet of German Creek, for a combined cost of $28,600 (*New Northwest* 1871f:3). The following year, 1872, Chinese companies purchased additional mining claims at German Gulch and nearby French Gulch, totaling $61,000 (Raymond 1873:272). For context, between 1871 and 1872, the Chinese purchased at least $106,600 of mining claims in the German Gulch area, translating to over $1.8 million in today's dollars (Williamson 2010). As in Blackfoot City, the Chinese purchasing claims in German Gulch must have possessed financial backing, likely through wealthy merchants and their connections to voluntary organizations. The

Figure 5: Selection of Chinese artifacts recovered from excavations at German Gulch. (Photo by author, 2008.)

purchases were indicative of a rapidly growing Chinese population in German Gulch, from only five in early 1870 to upward of 150 in July 1871, and 225 in June 1872 (*New Northwest* 1871f:3, 1872f:4). The shift in population in German Gulch led one reporter to claim that in 1876, "the Chinese have almost taken full possession of German Gulch" (*Helena Independent* 1876b:3). Dramatics notwithstanding, the reporter was accurate about the situation not only at German Gulch but also at various placer districts outside German Gulch or Cedar Creek.

In the early 1870s Chinese miners entered the German Gulch district as the easily worked placer deposits began to decline in productivity for the European American miners. By the late 1880s mining in German Gulch shifted to lode mining in the upper reaches of the drainage, and the majority of Chinese miners left the district. For a century, from the 1890s through 1990s, episodic yet intensive hard-rock mining significantly altered the physical landscape of upper German Gulch, while low-intensity placer mining did little to alter the rest of the drainage.

In the mid-1980s Pegasus Gold Inc. purchased all the lode claims at the head of German, with goal of creating a large-scale open-pit mine. With the need for widening the existing Forest Service road to accommodate large mine equipment, GCM, Inc., a contracting cultural resources firm from nearby Butte, surveyed the entire German Gulch drainage from its head to the confluence of German, Beefstraight, and Norton creeks. Pedestrian surveys discovered 54 areas of historic occupations, termed "localities," within the drainage (Fredlund et al. 1991:1). In 1988 and 1989 the company performed archaeological investigations of several areas within the six of the localities along the gulch. The majority of excavations focused on two areas directly impacted by the road widening, termed Lower and Upper Town, respectively. All tested localities had extensive Chinese archaeological remains, suggesting that for some period Chinese dominated the demographic and physical landscape, with the majority of European American miners working at the head of Upper German Gulch. The excavations discovered thousands of artifacts relating to the Chinese occupation of German Gulch and represent the most significant Chinese archaeological collection in Montana (Figure 5).

It is clear that the archaeological deposits at German Gulch relate directly to the Chinese domination of the area in the 1870s and into the 1880s. At all tested localities archaeologists encountered Chinese cultural materials. Of these artifacts some are mundane and common to nearly all Chinese sites in the American West, yet some, such as a sheephead fish and even half a coconut, provide some clues to the depth of German Gulch's connection to a world system of trade and commerce. The close access of German Gulch's Chinese population to ethnic import markets in Deer Lodge and Virginia City, and their proximity to the major trade routes coming from Utah and the trans-continental railroad, likely account for the diversity of this assemblage. In addition to German Gulch's proximity to such transportation opportunities, it is also probable that the Chinese store owners actively took advantage of these opportunities to provide more variety to their consumers than merchants did—or were able to do—at other encampments in Montana Territory. Whatever the case, it is remarkable to imagine a globular jar full of dried or pickled sheephead fish traveling by sailing vessel from China to San Francisco, then by rail to Corinne, Utah Territory, and finally being delivered over 400 miles away in German Gulch by freight wagons hauled by oxen, mules, or horses.

With Pegasus Gold filing for bankruptcy in 1998 after retrieving over $400 million in gold from the German Gulch operation, the future of German Gulch's archaeological features remains unclear. The retention pond holding back metric tons of contaminated water from the cyanide operations lies along a geologic fault line susceptible to tremors (*Montana Standard* 2009:1). The Pegasus bankruptcy effectively forces the State of Montana to handle the reclamation of the mine site, which might adversely affect several archaeological features relating to the Chinese heritage of German Gulch. If this is the case, there may be future CRM-related opportunities to return to the site to collect more archaeological data.

### *Blackfoot City*

Similar to German Gulch, Blackfoot City on Ophir Creek started out dominated by European American miners in the 1860s, but by the early 1870s mining opportunities attracted Chinese workers. As noted in the 1870 census, Chinese accounted for 22% (111 of 499) of Blackfoot City's population. In June 1870 Chinese workers from the Central Pacific Railroad began to enter the Blackfoot District, and reportedly, "the [Chinese] interpreter says one of the principal merchants is now on the road from San Francisco with $100,000 to invest in mining ground around Blackfoot" (*New Northwest* 1870b:3). In 1871 Chinese miners owned two miles of Ophir Creek below Blackfoot City, and a reporter observed that their intent to purchase more property confirmed that "they are truly called a curse to camp and country" (*Helena Daily Herald* 1871d:3). By the summer of 1872 estimates placed the total population of areas in and around Blackfoot City at 600, split evenly between Chinese and European Americans (*New Northwest* 1872e:3). The rapid settlement of Chinese in the Blackfoot City area led one reporter to state that "Blackfoot, as an American camp, is doomed and damned" (*New Northwest* 1870b:3). Some reports indicate that Blackfoot City had several Chinese stores, including a grocery that imported rice and other goods by way of freighters from Corinne to Helena and finally to Ophir Gulch (Wolle 1963:115).

### *Cave Gulch and Chinaman's Cove*

On the northern end of the modern Canyon Ferry Reservoir in the Big Belt Mountains, Cave and Chinaman's gulches contain archaeological habitations of Chinese workers and a palimpsest of mining features. A year after discovery of placer gold in Last Chance Gulch and nearby Confederate Gulch, non-Chinese prospectors hit paydirt in Cave Gulch in 1865 (Mineral Survey Notes 1877). Rossillon (2002:12) noted that the original discovery was in 1866, but this date should be taken back to 1865, as shown in the Mineral Survey Notes. Settlements of Cavetown, Springtown, and

Cashville existed from near the mouth of Cave Gulch along the Missouri River up the gulch for at least a few miles. Today most of Cavetown and possibly Springtown are located beneath the waters of the Canyon Ferry Reservoir created in the 1950s. Thousands of acres of forest lands in the Big Belt Mountains, including Cave Gulch, burned during a destructive series of fires in 2000. The Helena National Forest, which manages most of the land affected by the fire, hired Renewable Technologies Inc. (RTI) out of Butte to perform post-fire pedestrian survey and archaeological testing in 2001 and 2002. As part of their testing and recording regimen RTI located two archaeological sites in Cave Gulch with Chinese materials. In conjunction with the previously recorded remains of Chinaman's Cove, directly north of and adjacent to Cave Gulch, a complete picture of mining and Chinese occupation of this area emerges.

Similar to what happened in most other mining regions in Montana Territory, and the West in general, Chinese miners moved into the Cave Gulch vicinity during the early years of the 1870s after the production of the placers began to decline. As early as 1874 there were several dozen Chinese miners at work along Cave Gulch (Rossillon 2002:16) and probably also in Chinaman's Cove. The 1880 federal census listed 15 Chinese in Cave Gulch, 13 male miners, 1 male cook, and 1 woman "about town," suggesting prostitution. The next gulch north of Chinaman's Cove, Cooper Creek, had an additional 9 Chinese miners with 12 more miners and store owners north of Cooper in Oregon and York gulches. Taken as a whole, the pockets of Chinese miners on the northern end of Canyon Ferry made up a moderate population of nearly 40 (Federal Census Bureau 1880), and these counts likely underestimate the true numbers.

RTI's investigations of sites in Cave Gulch in 2001–2002 discovered two areas of Chinese occupation. Just within the boundaries of Forest Service land, a small settlement of several house pads and cellar features were the first Chinese artifacts encountered in RTI's excavations. This site is adjacent to Cave Creek, next to a mining claim originally located in 1865, abandoned in 1868, and located again in 1874 (Mineral Survey Notes 1877). In 1877 the claim adjacent to the site consisted of a 3000-foot-long bedrock flume and nine shafts with no mention of habitations. The presence of Chinese occupations and the associated cultural materials suggest that during the abandonment period, 1868 through 1874, Chinese miners worked the claims. It is also possible that the area encompassed by the site was a small satellite community associated with Cavetown, only a few hundred yards below the site. These interpretations of the site are distinctly different from those of RTI (Rossillon 2002, 2003), and are products of the author's reanalysis of the report and associated historical documents.

RTI recovered a number of Chinese artifacts from two building depressions and a trash scatter. The artifact assemblage indicates that Chinese miners occupied the two depression features sometime during the 1870s and 1880s. A distinct lack of temporally diagnostic artifacts at one location prevents full chronological interpretations. The most common Chinese artifact class found at these occupations included stoneware liquor jar fragments representing at least four vessels (Rossillon 2003:37, 43). The site's artifact assemblage is remarkably similar to that of Louiseville in Cedar Creek. The domination of Chinese stoneware, yet the absence of the thick-walled globular jars, suggests a small-scale occupation relying on easily portable goods. Of the three excavated features at this site, there were few European or American ceramic fragments, which is also reminiscent of Louiseville's assemblage.

The second Chinese site in Cave Gulch is over two miles upstream from Cavetown, and is possibly some type of water control station for mining operations in lower Cave and Chinaman's Gulch. According to Rossillon (2003:108), historical records indicate that residents considered Upper Cave Gulch as "Chinatown" during the late 1870s, probably alluding to the presence of Chinese miners owning most claims in the area. The area includes a number of house pads and associated features, and artifacts recovered from both excavations and surface collection indicate a series of occupations by both European American and Chinese populations.

In this case, the artifact assemblage is the not the only significant facet of the site. RTI recorded an earthen dam that extended across the bottom of Cave Gulch in 2002, only 450 yards from the Chinese and non-Chinese features. At this location Cave Gulch narrows to a rocky cut that allows for easy control of water for downstream operations. In addition, Rossillon (2002:47) notes that miners apparently lined the walls of the creek with stacked rock, providing better control of water flow and sufficient protection against flooding or damage. Heavy floods after the 2000 wildfire destroyed many of the features recorded by Rossillon (2002; 2003), and revisits in 2008 did not locate the earthen berm, retaining walls, or even some house pits near the creek. It appears that at least one or two feet of sediment have washed into the bottom of the gulch since 2000.

Chinaman's Cove Recreation Area is a significant site recorded in the

Figure 6: Chinaman's Cove showing hydraulic mining–created gully, photo facing southwest. (Photo by author, 2008.)

Cave Gulch vicinity that relates directly to Chinese occupation of the area, now managed by the Bureau of Reclamation. Chinaman's Cove consists almost completely of physical landscape features relating to mineral exploitation of the area; any historic occupation is now located under the waters of Canyon Ferry Reservoir. In the late 1860s through the 1870s miners, Chinese and non-Chinese, used hydraulic mining equipment to blast away the gold-bearing gravels along Chinaman's Gulch (Figure 6). While there is no supporting historic documentation to indicate a Chinese operation at this location, the presence of a Chinese coin discovered on-site just below the water line several years ago is suggestive of some association. With Chinese companies working in Cooper and Cave Gulch, on the northwest and east sides Chinaman's Cove, respectively, the likelihood that they were working within the site boundary itself seems logical. Chinaman's Cove is a current recreation area with campsites, toilets, and boat launch set within the scars left by hydraulic mining. Other than a single length of metal hydraulic mining pipe and the Chinese coin, the site is devoid of artifacts, though it is probable that historic occupation now lies at least ten meters underwater if not washed out and destroyed by the mining efforts.

Whereas German Gulch provides an excellent example of a placer mining archaeological landscape, Chinaman's Cove contains all the components of a hydraulic mining landscape. Aerial photographs and surface surveys reveal the location of at least four ditches entering Chinaman's Cove area from the northeast or Cave Gulch side of the ridge. In addition, at the end of the ridge overlooking the east side of Chinaman's Cove, there is a large water retention dam constructed of an earthen berm. Hydraulic mining required substantial amounts of water and a large drop, or "head," to create force enough to blast out of a monitor or "giant" nozzle to wash the gravels. A retention dam at the top of the ridge provides a significant pool to tap and then divert to hydraulic operations on several flanks of Chinaman's Cove. As gold-bearing placers existed in the center and mouth of Chinaman's Gulch, the ridges never experienced mining and are one of the best-preserved hydraulic mining landscapes in Montana.

In order to understand fully the story of the Chinese in Cave Gulch, and their mining operations, a complete look at the landscape is necessary. Starting at Chinaman's Cove, which is the end result of the mining operations, and moving north, there is an area managed by the Bureau of Land Management and the Helena National Forest exhibiting significant placer and hydraulic mining scars. In this area and extending onto the ridge separating Cooper and Cave Gulch, at the head of the shorter Chinaman's Gulch, there are piles of placer tailings, ditches, and more earthen dams. The water flowing into Chinaman's Cove passed through this area before descending. Just below and east of the ridge is the Chinese miners' encampment along Cave Gulch, which possibly housed the laborers who constructed or maintained the ditches and dams. Finally, the upper site sits at a point where ditches connected to Cave Creek and brought necessary water to hydraulic or placer operations in Chinaman's Cove and lower Cave Gulch. While the artifact assemblages of the Chinese sites in the Cave Gulch area are important, the primary significance of these three sites is their illustration of an entire mining system for the exploitation and modification of the physical landscape.

### *Last Chance, Pike's Peak, Big Prickly Pear, Garnet Range, and Old Lincoln*

As promising placer mining deposits waned, other districts shared histories similar to those of German and Ophir Gulch, with European American miners moving on while the Chinese lingered to carve out a living.

For example, by the summer of 1872, Chinese companies were the only mining groups working the lower parts of Last Chance Gulch in Helena (Raymond 1873:282). In some districts, like Pike's Peak northwest of Deer Lodge, Chinese were the only crews at work in the drainages, and Raymond (1873:274) notes that "wages in this place are about $50 per month [and] they are exclusively employed by their own countrymen." By 1871 Chinese miners represented the bulk of the population working on the Big Prickly Pear mines near Montana City, and by 1872 the area was devoid of all but a handful of European American miners (*Helena Daily Herald* 1871h:3; Raymond 1873:288). In Bannack, which had long since passed from the front pages of Montana Territory's newspapers, three Chinese companies had about 50 men working below town along Grasshopper Creek (Raymond 1873:262). Declining districts in the Garnet Range, east of Missoula, began a demographic transition to almost exclusively Chinese occupations of placer camps in the mid-1870s, as evidenced by the town of Yreka in 1876: "the China town is larger than the white burg" (*Helena Independent* 1876a:3). In contrast, some mining districts like Old Lincoln (Didriksen and Didriksen 1994) had a small Chinese population in the mid-1870s but never coalesced into a substantial population, as in other areas.

### *Yogo Town and the End of an Era*

The last major placer gold rush in Montana Territory occurred in 1878 along Yogo Creek in the Little Belt Mountains southeast of Great Falls. With news of the strike thousands of European American miners stampeded to the workings along Yogo Creek, though by 1880 the area was largely abandoned (Voynick 1985:10). It is unclear how many Chinese also came to the area in the rush, although it is possible to extrapolate from the previous patterns that at least a few dozen likely attempted to work claims. The 1880 federal census noted two Chinese laundrymen in Yogo Town, significant because the town had a population of only a few dozen by this time (Federal Census Bureau 1880). Archaeologists recorded the remnants of Yogo Town in the early 1980s. During surveys in 2008 archaeologists discovered a hand-stacked rock hearth similar in size and shape to those discovered at China Gulch and Mary Anne Gulch. Excavations of the hearth and associated dugout at Yogo Town in the summer of 2009 recovered no ethnic markers associated with the Chinese (Merritt and Mogstad 2009:50).

Yogo Town's rapid boom and bust was indicative of a major shift that had occurred in Montana Territory during the 1870s. Consolidation of existing claims and intensification of mining works of previously placer-worked gulches and streams was the perpetual pattern in 1870s Montana Territory for both European American and Chinese miners. By the latter part of the 1870s miners had removed most of the easily recovered placer gold deposits throughout Montana Territory. Employment of new techniques such as hydraulic mining as early as the 1860s allowed further exploitation of many mining districts where superficial deposits had been played out by earlier miners. Chinese found widespread employment in hydraulic operations owned by European Americans or their own companies (Rohe 1996). Chinese miners worked hydraulic operations in Confederate and Cave gulches east of Helena, and along Elk, Bear, Gold, and Pike's Peak creeks northwest of Deer Lodge; they were also involved in widespread hydraulicking along Alder Creek near Virginia City. Dredging and other remining after 1900 destroyed most physical evidence of Chinese hydraulicking, except in German and Cave gulches. The transition to these heavily capitalized ventures in the 1870s began to force Chinese miners out of Montana's mining districts and into more service-based industries, although mining capitalization and transition were not the only factors affecting the economic livelihood of Montana's earliest Chinese residents.

## Montana Territory's Legal Attack on the Chinese

The first legal attempt specifically to exclude the Chinese was fueled by competition in the laundry business in the mid-1860s. Reacting to the construction of several Chinese laundries in downtown Helena, a group of European American washerwomen purchased editorial space calling for the boycott of those businesses in 1866. A few weeks after the call for boycott, a group of Chinese laundrymen wrote an editorial stating that they were "Good Chinamen" and asking "that the good people of Montana may let us earn an honest living by the sweat of our brow" (*Montana Radiator* 1866:1, cited in Wunder 1980:20). It appears that the initial calls for boycott fell on deaf ears in 1866, but in 1867 a similar dispute occurred in Diamond City, 40 miles east of Helena (Flynn 2006:108; Stoner 2007:30). Charlie Lee, a Chinese man, constructed the first laundry in Diamond City in 1866, and by the mid-1870s three Chinese laundries operated in the town (Flynn 2006:53; Stoner 2007:31). Growing negative sentiment toward Chinese laundries led to a restrictive tax levied on all "male-owned" establishments of that kind in Montana Territory. The initial tax provision specifically stated that the law applied only to Chinese laundries, but the territorial supreme court struck down the law as it focused on one ethnic group. A reworked tax provision that changed the wording to "male-owned" laundries passed the territorial government in 1869 and remained on the books into the 20th century (Spence 1975:199–200; Wunder 1980:26).

Over 25% of the Chinese laundry operator's budget was subsumed to this tax, though the enforcement and collection it seems to have been relatively variable (Bernstein 1999:23). In two court cases in the 1890s the state supreme court upheld the laundry tax as constitutional on the basis that the Chinese litigants were not American citizens and that the act was a fee and thus not required to apply equally to all ethnic groups (Wunder 1980:26). Wunder (1980:27) notes that one possible reason for the supreme court's opinion was the potential financial loss to the state if they judged the tax unconstitutional. Restrictive and harmful as the tax seemed to be for Chinese operators, the number of Chinese laundries nevertheless grew rapidly throughout the 1870s and 1880s. Effectively the laundry tax ceased being enforced after 1913, when Quong Wing of Helena won a case at the district level but lost in the supreme court (Goldstein 1988:96). Even with the act upheld by the supreme court, the State of Montana ceased collection of taxes on laundries in the late 1910s (Wynne 1978:63).

After Chinese companies and individuals began to purchase placer ground in nearly every mining district in Montana Territory during the early 1870s, public sentiment shifted toward exclusion of this population by means of legal restrictions. More than one newspaper report during this period noted that the purchase of placer ground by Chinese miners was throwing American citizens out of employment, even though federal mining reports suggested otherwise (Raymond 1869:141, 1870:260, 1873:292). Indeed, it appears that the territorial legislation perpetrated injustice against the Chinese, not the reverse.

For instance, in November 1871 two Chinese individuals purchased lode claims in the Elkhorn Mountains southeast of Helena. Ah Chong and Ah Song purchased placer and lode claims from W. D. McIntyre and V. Simmons for $3,000 in 1871. This property was located on the "main branch of Crow Creek, also known as Wilson Creek," located in the "Warren Mining District" in an area that would become known as Ruddville (Jefferson County Courthouse 1871:247). Further research of the mining claim records indicated that on September 25, 1871, V. Simmons had purchased and consolidated all the claims in Wilson Gulch from James L. Wood, A. Campbell, A. J. Sanders, Robert Holcomb, Charles Huse, and Joseph Harvey for a sum of only $100 (Jefferson County Courthouse 1871:199). Then less than two months later, in partnership with W. D. McIntyre, he resold all the claims in bulk to the aforementioned Chinese for $3,000, a significant profit. This pattern was repeated countless times across the state. Even though the European American majority profited from the Chinese, they continued to rail against those living and working in Montana Territory's mining districts. Boswell (1986) best describes the situation in noting that when the economy contracts, as it did in the 1870s, there is a significant increase in discrimination. Anti-Chinese sentiment led to major attempts to subject Chinese to exclusionary laws on local, state, and federal levels.

The 1868 Burlingame Treaty between the governments of China and the United States promoted the immigration of thousands of Chinese laborers to America to fill the need for cheap labor in various industries, including manufacturing and the railroads. Additional provisions within the Burlingame Treaty prohibited the naturalization of Chinese for American citizenship (U.S. Department of State 1902:114). This legal precedent would have serious repercussions for the Chinese residents of the United States and would play out in Montana Territory's legislature and court system in the 1870s.

During December 1871 Seth Bullock, a member of the Montana Territorial Legislature from Lewis and Clark County, introduced a rough bill that would "exclude all Chinamen from acquiring any right, title or interest in mineral lands, by discovery, possession, purchase or lease, within the limits of the Territory" (*Missoula and Cedar Creek Pioneer* 1871b:2). Bullock's proposed bill was nearly the same in language as bills already enacted in California and Idaho and remarkably similar to those proposed in North Dakota, Oklahoma, and South Dakota around the same time (Ross 1909:176). The bill followed on the heels of miners' meetings in a handful of Montana's mining districts that sought to exclude the Chinese from ownership of claims and any form of real property, though no sanctions were ever successfully passed. For example, miners in Cedar Creek Mining District in the western part of Montana Territory failed to pass a resolution expelling the Chinese from the area. Most opponents of the

Cedar Creek measure stated that they had sold equipment and goods on credit to the Chinese, and expelling the population would bankrupt local merchants (*Missoula and Cedar Creek Pioneer* 1870b:3, 1870c:2).

In late December 1871, Bullock formally introduced a bill to the upper house of the territorial legislature under the title "To Prevent Chinese from Acquiring Mining Ground in Montana." With the submission of this proposal a surprisingly vigorous debate, given the prevailing attitudes toward Chinese labor in the United States, erupted between rival newspapers and editors. The editorial board of the *Helena Daily Herald* fully supported the bill and went even further in providing some recommendations for revisions that they hoped would garner better support from the various political factions. Specifically, in an editorial on December 27, 1871, the *Helena Daily Herald* first berates the *New Northwest* out of Deer Lodge for its "inability to discern" the necessity for such a bill but also suggests that Bullock and supporters should alter the bill's language (*Helena Daily Herald* 1871j:3). As it stood, Bullock's bill focused purely on the Chinese. It originally stated that "from and after the passage of this act, no subject of the Empire of China shall acquire or hold any real estate, placer or quartz mines." The *Helena Daily Herald* editorial staff suggested a broader application with an amended section: "That from and after the passage of this act, no subject of the Empire of China, the Empire of Japan, or other alien, who, by laws or treaties of the United States, are prohibited from becoming citizens thereof, shall acquire, possess, or hold any interest in or to real estate, mining claims" (*Helena Daily Herald* 1871e:3). Thus the proposed new language more closely followed the stipulations of the Burlingame Treaty of 1868 in reference to the Chinese, but it also provided future protection of real property from sale to any alien party.

In January 1872 W. W. Dixon of Deer Lodge introduced a modified version of Bullock's bill to the House of Montana Territory. The second version of the bill was slightly different than Bullock's original with the main changes including: (1) amending language to prohibit all aliens, not just Chinese, from owning property, following the editorial suggestion by the *Helena Daily Herald*; and (2) the only property specifically addressed for prohibition of ownership by aliens was placer claims. In a telling editorial on the proposed bill, the *Missoula and Cedar Creek Pioneer* (1872a:2) claimed that the provisions of the bill were a humanitarian endeavor to protect the Chinese from the pervasive violence in Montana's mining camps:

*In the interest of humanity, order and law, Mr. Dixon takes the earliest opportunity to allay this feeling of discontent and to protect the Chinamen from mob violence, by introducing this bill, which protects the citizen and those who may become such in their "rightful heritage," and removes the cause which has led to mob violence and summary execution of Chinamen.*

In addition, the same editorial stated that the proposed bill would effectively force the Chinese to leave Montana for good, and congratulated the Montana Territorial Legislature for its "bloodless solution" to the Chinese question. Finally, under the bill's provisions, any placer claims held or purchased after passage of the act would be forfeited to the Territory of Montana to be resold at cost to legal citizens of the United States.

After some apparent change in opinion since the 1871 editorial by the *Helena Daily Herald*, the editors of the *New Northwest* (1872b:4) issued a statement in 1872 that explicitly supported the Alien Bill or Law, as it was now termed, and furthered public opinion on the matter:

*The placers are the filled coffers of wealth to sustain Montana till her acres are tilled. We have been permitting them to be robbed by those who have, and will have, no interest in common with "Republican America." We desire that this shall cease. As for the Chinaman, we bear him no malice. As he got into Montana so let him get out.*

In the same newspaper edition as this editorial rallying support for the Alien Bill, a response letter voiced a dissenting view on the issue. A letter by "Ursa Minor," a pseudonym of some unknown author in Little Bear Gulch, disparaged the attitude of expelling the Chinese and instead saw this as an opportunity for good Christian Montanans to embrace these individuals. Ursa Minor specifically pointed to the opportunity for educating the Chinese currently in Montana in American governmental and religious ideals and thus spreading ideas of democracy back to China. Ursa Minor believed that it would be better if all the European Americans in Montana were forced out of placer mining enterprises and instead focused their attention on farming, as "that one dollar made tilling the soil is worth ten dollars made at mining" (*New Northwest* 1872c:4). Ultimately, there were few opponents to the Alien Law in the government, and the measure successfully passed the Montana Territorial Legislature in January 1872.

There is but a trace of evidence on how the Alien Law played out in

Montana Territory during the first few years. A report by Rossiter W. Raymond, president of the American Institute of Mining Engineers in 1873, noted that he could locate only "two or three instances of actually perpetrated injustice" toward the Chinese as a result of the Alien Law (Raymond 1873:292). At least one newspaper account mentions that Chinese companies of German Gulch were nervous to purchase more mining claims, "unless [the Chinese] can be guaranteed uninterrupted possession, which no one is willing to do at present" (*New Northwest* 1872h:3). Another example of the impact of the Alien Law comes from Horse Prairie, where two Chinese miners who purchased $70,000 worth of mining ground from the Yearian brothers reportedly filed papers to become American citizens to avoid the law's penalties (*New Northwest* 1872k:3). Even with the passage of the Alien Law and the apparently small effect it had on Chinese purchasing habits, some in the press wanted still more restrictions. According to those wanting more stringent restrictions, the action was required so that "we can so completely despoil them in one year's time that they will curse the country, shake the dust from their coattails, and leave for 'Their own, their native land'" (*New Northwest* 1872k:3).

In the report by Rossiter W. Raymond (1873:272), supplied to the secretary of the treasury for the United States, he expressed his disgust, calling the bill, "outrageous," a "ludicrous illustration of human stupidity," and a "rank dishonesty." The crux of Raymond's opposition to the bill was that it seemed a folly to him that European American miners would want to exclude Chinese from purchasing what he saw as second-class or previously worked claims. Raymond viewed the exclusion of Chinese from purchasing claims as a position that would hurt not only the Chinese but also the European American miners who were trying to offload low-paying claims. Finally, Raymond offered a warning about the potential abuse of Chinese by unscrupulous European American individuals who would "by virtue of laws of their own manufacture, excluding foreigners from rights of original location, get the land in the beginning for nothing, and then, after having skimmed it of its richest treasure, sell it at their own price [to the Chinese], and pocket the money of the purchaser," and then potentially revoke the title under the provisions of the Alien Law and be able to resell the land yet again for profit (Raymond 1873:292).

In 1874 Montana territorial District Attorneys Henry N. Blake and J. C. Robinson lodged a formal action under the Alien Law against Fauk Lee, a Chinese mine owner from Deer Lodge County, who had recently purchased

3,000 feet of placer claims from a European American individual (Blake and Hedges 1877:126). The lawsuit was potentially a collusion between the district attorneys and Chinese parties to test the legality of the Alien Law, as Blake had defended Chinese clients before (Wunder 1980:24). Regardless, Blake and Robinson successfully argued their case at the lower court level in Montana, with the result that "judgment was rendered for the plaintiff, that the mining ground described in the complaint be forfeited to the Territory, and sold in pursuance of the provisions of the [Alien Law]" (Blake and Hedges 1877:126–127). It is a possibility that this lawsuit against Fauk Lee was a direct result of actions by leaders of the Chinese Six Companies in San Francisco, who in May 1873 instructed their subordinates in Pioneer, and probably other mining camps in Montana, "to buy all the mining ground they can at whatever figures they consider it worth, and the San Francisco head men will contest the Montana Alien Law if their title is disputed" (*New Northwest* 1873a:4). The full extent of Six Companies' involvement in the Fauk Lee case is undetermined, though the mining property at stake was a short distance from Pioneer. There is a distinct yet indeterminate possibility, as suggested by Wunder (1980:24), that the Chinese Six Companies or some other Chinese party paid Blake to lodge the lawsuit against Fauk Lee so that they could legally challenge the Alien Law in court.

After losing their fight in the district court, Fauk Lee and his attorneys filed an appeal to the Montana Territorial Supreme Court, and the final verdict was rendered by Judge Knowles. Specifically, the territorial supreme court justices considered: (1) the rights of aliens within the territory; (2) whether the territorial legislature had the authority to create such law; and (3) whether the Alien Law conflicted with the Organic Act that created Montana Territory in 1864 (Blake and Hedges 1877:127). Under the analysis of alien rights, the supreme court's majority opinion found that under the Burlingame Treaty with China, Chinese individual rights as persons from a most-favored country protected their ownership of such claims (Blake and Hedges 1877:129). Second, the court found that the territory did not have the authority to enact the Alien Law in the first place, as the Organic Act of Montana Territory provided no specific sovereignty rights that would be necessary to enact legislation affecting aliens to the United States as a whole (Blake and Hedges 1877:134; Wunder 1980:24). Finally, the supreme court's opinion stated that the territorial Alien Law was in conflict with the 1872 Hard Rock Mining Law that the United States government had passed, affecting all mineral lands in the country (Blake and Hedges 1877:138). This finding successfully destroyed the effectiveness of Montana Territory's Alien Law and protected one of the main economic opportunities for the Chinese. The success of the appeal by Fauk Lee is truly remarkable, as the finding occurred during one of the most severe economic depressions in American history (Spence 1975:34), which led to even further fomenting of anti-Chinese sentiment among the working classes and efforts to further exclude them from the workforce (McLain 1994:96).

Even though Fauk Lee successfully defended his and all Chinese rights to owning mineral claims in Montana, the federal 1872 Hard Rock Mining Law effectively kept Chinese ownership off the official books. According to Hartman (1992:142), the 1872 Hard Rock law's "proscription against location of claims by aliens is enforceable only at the instance of the United States." Thus on a local level mining claims and deeds could be purchased and sold by aliens, in this case Chinese, without fear of private parties attempting to challenge valid ownership. The restrictions in the law allowed ownership of unpatented claims by the Chinese, but never patented claims, as the United States would then challenge the rights of the alien claimant (Hartman 1992:142). This legal reality led to omission of hundreds of Chinese claims from being included in official records, creating even more holes in the documentary record of the Chinese in Montana.

## Violence and Death in the Montana Territory Chinese Communities

Death and violence in the Chinese communities of Montana Territory were no more severe or different from patterns evidenced in the European American majority. This noted, there was a difference in the manner Chinese immigrants dealt with violent and domestic problems within the community. As observed by Raymond (1869:140), "their mutual quarrels seldom or never require arbitrament [sic] of the courts." The Chinese method of self-regulation of disputes was unique in Montana Territory and other areas of America. It appears that most self-regulation occurred in domestic disputes over money or nonfatal confrontations between Chinese individuals or companies. An 1867 incident in Helena illustrates the self-regulation of the Chinese population by a type of group consensus, though the true inner workings of the trial system remain unclear.

On the night of November 20, 1867, Na Lung supposedly doused an old shirt with coal oil and placed the incendiary in a wood pile behind the

Jing, Sing & Company store on Cutler Street (*Helena Daily Herald* 1867a:2). Apparently Na Lung was immediately suspected of the attempted arson due to community knowledge of an existing grudge toward the owners of Jing, Sing & Company, and the Chinese community convened a type of anonymous trial by jury the following Sunday. During this trial Chinese community members anonymously placed in a box slips of paper with the name of the person they believed started the fire, and the votes were tallied. Voters convicted Na Lung of arson and sentenced him to hang at the earliest opportunity. It is not known what happened to Na Lung, as he apparently disappeared, and Jing Sing & Company offered a $500 reward for his capture (*Helena Daily Herald* 1867b:2).

In the period between 1862 and 1880 there were several acts of violence, murder, and dishonesty within the Chinese community, though few spilled over into Montana Territory's legal system. Violence against the Chinese in Montana Territory was similar to that in other states and territories in the United States and included lynching and random acts of violence. Some of the violence toward the Chinese stemmed from vigilante justice, which pervaded territorial Montana in the 1860s and 1870s (i.e., Dimsdale 1977; Allen 2004; Thompson and Owens 2004).

For instance, the first and only known lynching in the mining community of Henderson, north of Philipsburg, and the first recorded in Granite County, was of a Chinese individual accused of stealing $400 in 1871. According to accounts, the Chinese man broke into the store of Caplice & Company and robbed the till. Miners arrested the Chinese individual during his exit from the store, and a citizens' trial found him guilty and sentenced him to hang by a vote of 19 to 18. The sentence was carried out immediately, and though the newspaper editorial condemned the hanging and called for an investigation, no further mention of the incident appeared in newspaper accounts (*New Northwest* 1871h:3). In Beartown, a Chinese man was hung, or "choked" as euphemistically reported by the *New Northwest* (1874e:3), for the reported theft of 350 pounds of flour from a store. Even after the preceding lynching the newspaper reporter noted that the Chinese "immunity from punishment is justly looked upon as a great calamity to the expansion of good morals, and is only excusable on the ground of a scarcity of cheap labor in Beartown" (*New Northwest* 1874e:3). In another instance of apparent vigilante justice, hunters discovered a Chinese man hanging from a tree two miles east of Helena. In this case investigators did not locate a motive or perpetrators, and it appeared the body had been hanging for several weeks or months (*Helena Daily Herald* 1871i:4).

The most notable violent encounter between Chinese and European Americans in Cedar Creek was during 1872, when someone broke into a store in Forest City, two miles above Louiseville, and stole a safe containing $4,000 dollars in gold (*Missoula and Cedar Creek Pioneer* 1872b:2). Four Chinese men were immediately suspected of the crime, and a posse under Sheriff Ed Warren began to pursue the individuals. During the initial excitement a mob burst into the cabin of a Chinese couple on the outskirts of town:

> *They attempted to make the couple confess by putting ropes around their necks and running the ropes over a rafter in the cabin, pulling them up and letting them down. After nearly choking the two to death the mob decided they knew nothing of the robbery and shamefacedly left them alone. Later no one liked to admit any part in the incident. (Hahn 1986:31)*

A few weeks later Ed Warren and the two posse members returned to Forest City with the money and a tale. At this point the story diverges into several scenarios, each reported as fact, though all different in substance. According to the initial reports, Ed Warren and his posse encountered the Chinese men on a trail heading through the Coeur d'Alene Mountains, approximately 40 miles northwest of Cedar Creek. The story continued that the Chinese refused to surrender, at which point the posse fired and wounded several of the Chinese, who still managed to escape with the money (*Missoula and Cedar Creek Pioneer* 1872c:2). A second version states that Ed Warren and crew encountered the Chinese in their camp, surprising them while nude and washing their clothes. In this version, perhaps concocted for its humorous and humane nature, the Chinese ran into the woods unclothed, leaving the money behind (*Missoula and Cedar Creek Pioneer* 1872d:2). Another version noted that there were only two Chinese men involved with the robbery, and Warren shot and killed one and arrested the second (*New Northwest* 1872g:4). A few months after the incident, and with no other party locating the fugitive Chinese thieves, the guilty conscience of a posse member leaked that Ed Warren and others murdered the Chinese during the arrest and dumped their bodies into the Clark Fork or DeBorgia River. Later a second party found the badly decomposed bodies on the riverbank and burned the remains over a large bonfire (Hahn 1986:31).

Another instance of violence against the Chinese in Montana Territory occurred in 1873 in Rocker City near Butte and received far more press and legal condemnation than the other incidents. In early July 1873 two European American men living on Silver Bow Creek, just outside Rocker, supposedly believed that a Chinese man had stolen a ring and watch from their cabin. The two men, Daniel Haffey and John McLaughlin, captured the supposed thief and enacted their version of vigilante justice (*New Northwest* 1873d:3). Another source mentioned that Haffey and McLaughlin went into Butte and Rocker City boasting of the deed (Warren 1896:69). A newspaper report carried the gory details; "when found his feet were on the ground, and a heavy blow on his head with a six-shooter or some other heavy weapon indicated that he was struck after he was strung up and, swooning away, strangled to death" (*New Northwest* 1873d:3). After multiple attempts by both European Americans and Chinese living in the area to cut down the body met with threats of violence from Haffey and McLaughlin, word was sent to Sheriff Warren of Deer Lodge. When Sheriff Warren arrived at the location the Chinese body was still hanging from the noose, and both Haffey and McLaughlin had fled the scene. On one grisly note, Sheriff Warren discovered that the eyes of the murdered Chinese man were gone due to decomposition and the work of flies. The notation of the missing eyes is important, as a later report stated that no Chinese miners wished to return to the area as "the dead man is looking at them," possibly an expression of traditional superstition (*New Northwest* 1873e:3).

Newspapers condemned the murder, and a coroner's inquest proved that the Chinese man did not steal any of the objects Haffey and McLaughlin claimed he had taken. Sheriff Warren set a $100 reward for the capture of the two men, and within a week citizens captured McLaughlin attempting to leave the county (*New Northwest* 1873d:3). Under further review, Sheriff Warren determined that McLaughlin was the same man who, under the alias of William Terry, had accosted a Chinese doctor and broken out windows of a Chinese store in Deer Lodge one month before the hanging incident. In the earlier incident it appears that a drunken McLaughlin accosted a Chinese doctor and merchant and then broke windows of the Chinese store with his fists and with rocks before being arrested. On May 31, 1873, the judge sentenced McLaughlin, who lived in Bannack at the time under the alias of William Terry, to $25 and costs of court and one month in jail.

It would appear that only a few days after his release from a one-month jail sentence he took his possible resentment out on the first Chinese individual he came across, with support from Haffey (*New Northwest* 1873b:3). It is interesting to note that the *New Northwest* reporter handled the initial assault and battery in Deer Lodge in May as more of a joke than an unfortunate crime, but that tone changed after the murder. For example, after the murderous lynching the paper stated that the Chinese man was "entitled under our laws and the humanities of a civilized people to the same protection in life and property as though his skin had been white and his God our God, and his murder is a stain and humiliation upon all Montana" (*New Northwest* 1873d:3). Even in the face of this crime, the reporter had to couch condemnation with: "We wish there was not a Chinaman in Montana; believe them to be a blight upon the country, but they are here by lawful authority" (*New Northwest* 1873d:3).

There was never any mention in historical accounts of the capture of Daniel Haffey, as John McLaughlin (a.k.a. William Terry or John Collins) was the only person tried for the crime. Judge Knowles of Deer Lodge found John Collins guilty of murder in the second degree and sentenced him to life imprisonment, the first such verdict in Montana territorial history, though prosecutors pushed for the death penalty. Again reporters expressing their agreement with the punishment could not pass up the opportunity to denigrate the victim of the crime: Chinese people, "however objectionable as an element of population, are still under the protection of our laws, entitled to every right of person and property that we are, and who by very reason of their helplessness and inferiority should be exempt from personal violence" (*New Northwest* 1873f:4). Even in murder situations, reporters and the European American citizens of Montana could not relinquish their own racial bias against the Chinese. Nine years later, in 1882, Judge Knowles granted John Collins a full pardon due to a petition by 200 Deer Lodge County citizens and two jurors from the original murder trial (one of whom was the prison warden in 1882). Judge Knowles reanalyzed the case, believed the claim that Collins was under the influence of alcohol and remembered nothing of the crime, and thus released him on the stipulation that "he would forever abstain from the use of intoxicating liquors" (*Helena Independent* 1882c:2). Collins, upon release, went to San Francisco and immediately went to sea, ending one of the most unfortunate chapters in Montana Chinese history.

The last major violent episode between Chinese and European American Montanans during the initial period occurred at Blackfoot City in 1871, and the resulting trial had repercussions across the territory and the West. Blackfoot City, known to the Chinese residents as "entrance to the valley or where the valley begins" (Avon Get Together Club 1977:411, cited in Foster 1992:37), had a substantial population of both European Americans and Chinese during the early 1870s. During March 1871 a significant clash occurred between a European American miner, John Martell, and nearly all the Chinese mining companies working for three miles below Blackfoot City. The dispute began apparently over access to water flowing through a ditch constructed in 1866 from Ophir Creek downstream to the workings of Martell and his partner Kimerly (*New Northwest* 1871a:3). While the exact location of Martell and Kimerly's claim is not included in the report, the distance, three miles below Blackfoot City, would indicate that they were working near the confluence of Ophir and Carpenter creeks.

From the reports it appears that Martell hiked up Ophir Gulch to the location of the dam for the ditch, which was on a claim owned by a Chinese company; and apparently an altercation ensued in which one Chinese man was shot and killed. Since the Chinese company had purchased the claim from a European American miner the previous year, there was an ongoing dispute regarding who owned the "prior right" to the water that precipitated the affair (*New Northwest* 1871a:3). It is unclear who the aggressor was in the altercation, but Martell fired his pistol and instantly killed Ah Sing, foreman of the Chinese company, with a single shot to the head. Martell contended that the Chinese miners were planning on breaching the ditch, and drew their weapons to scare him away, at which time he defended himself, killing Ah Sing (*New Northwest* 1871a:3; *Helena Daily Herald* 1871a:3). After firing several more shots Martell took off on foot toward Blackfoot City, a few more miles up the gulch. The Chinese company, possibly totaling a dozen men, chased Martell, firing as they pursued him (*Helena Daily Herald* 1871a:3). A second Chinese company, which owned the next claim upstream from where the altercation occurred, heard the commotion and joined the pursuit of Martell (*New Northwest* 1871a:3). By the time Martell reached Blackfoot City, divested of his work boots during the pursuit, the estimated 20 Chinese miners had fired 30–40 rounds. The sheriff in Blackfoot City arrested not only

Martell but also 10–12 Chinese who had followed him into town (*New Northwest* 1871a:3).

At an initial hearing the next day the judge refused bail for Martell while releasing the Chinese prisoners on a $1,500 bond, which caused "general indignation on the part of the whites" of Blackfoot City (*Helena Daily Herald* 1871b:3). To the reporters, it seemed most egregious that refusal of bail was based only on "the fact that none other than Chinese evidence was adduced against him" (*Helena Daily Herald* 1871b:3). It appeared that the single most problematic part of this whole episode in Montana territorial history was the lack of trust placed in Chinese testimony by European American Montanans. During jury selection for the manslaughter trial of Martell, attorneys interviewed over 100 European American men as potential jurors, but none stated that they would believe the testimony of a Chinese witness (*Helena Daily Herald* 1871f:3; 1871g:3).

This being a major public trial of its day, the defense team consisted of high profile attorneys William Claggett, a future Montana Territorial Supreme Court justice, and William Dickson. The prosecution, in contrast, consisted of District Attorney McMurtry, Colonel Brown and Colonel Chadwick, all notable names in early Montana territorial history. All prosecution witnesses being Chinese, the court used as an interpreter a former sailor aboard ships trading with China, J. H. Oliver of Blackfoot City, due to his basic knowledge of the foreign language (*Helena Daily Herald* 1871f:3). The trial lasted roughly a month, which was longer than most, due to the problems securing jurors and the slow interpretation of testimony. On May 18, 1871, Judge Knowles acquitted John Martell of all charges. While most reports supported the decision, an editorial in the *Missoula and Cedar Creek Pioneer* stated their "inference that the people of Deer Lodge approve of shooting Chinaman for a pastime" (*Missoula and Cedar Creek Pioneer* 1871a:2). The position of the *Missoula and Cedar Creek Pioneer* editorial staff was more likely dependent upon a political rivalry with the *New Northwest* and *Helena Independent* than on the welfare and support of the Chinese population of Montana Territory, though it is hard to translate the argument fully. The true legacy of the Martell trial lies in the admission of Chinese testimony against a European American defendant. Precedents in California declared that Chinese could not testify against European American men, similar to the antebellum laws of the American South regarding African Americans. Specifically, the *New Northwest* ran a story detailing the robbery of Chinese miners by European American men in California, but as Chinese could not provide testimony the court dismissed the case (*New Northwest* 1871e:3).

### Chinese Women

While few primary sources and scholarly studies have focused on the Chinese experience in Montana Territory during the 1860s and 1870s, there is even less discussion of the female component of that population. In 1870 the federal census enumerated 128 Chinese women, largely located in Deer Lodge and Lewis and Clark counties. Of these 128 Chinese women the majority, 57, kept house, 53 engaged in prostitution, and the remaining 17 fell into various occupational categories. By the time of the 1880 federal census the female population of Montana Territory's Chinese fell to 78. The large disparity between male and female Chinese ratios (14.23 males to each female in 1870 and 20.51 males to each female in 1880) had severe widespread repercussions in the immigrant community. In comparison, the sex ratio for the entire population of Montana Territory was 4.2 males per female in 1870 and 2.5 males per female in 1880.

As noted by Hsu (2003), the largely male Chinese populations created a homosocial community where the primary relationships formed between men. While there is no clear evidence for homosexual activity among the Montana Territory Chinese in the 19th and early 20th centuries, the limited options for female companionship possibly produced such arrangements. Hsu (2003:232) also correctly notes that the bachelor communities of Chinese in the United States created further opposition to the mainstream norms of 19th-century Victorian or Progressive America, which focused on family. Thus the restrictive immigration laws of the United States created communities of Chinese that were even further from normality, which fueled restrictions and a perpetual cycle of otherness. Chinese men generally performed tasks associated with female gendered roles such as laundry, cooking, and domestic servitude, further polarizing European American perspectives of Chinese distinctions from American norms.

Most of the stories regarding Chinese women in Montana Territory during this initial period focused on prostitution, murder, and struggles over marriage. The first significant mention of women in Montana Territory's communities revolves around a murder mystery in Helena during 1867. In early September 1867, investigators found China Mary, an apparently notorious Chinese prostitute in Helena's Chinatown, dead in her cabin on corner of Water and Cutler streets. At the time of her murder China Mary was preparing to leave Helena for California with approximately $1,500 in gold and currency. She was heading to California with her fiancé, an African American male unnamed in reports.

On an early Sunday morning the unnamed African American man went to China Mary's cabin to help with preparations for the trip, which was scheduled to begin that morning. When he arrived at the cabin he found China Mary engaged in conversation with another man and left the area to return later. Upon returning to the cabin a few hours later he discovered China Mary dead on the floor, with her underwear twisted around her neck. A coroner's inquest indicated that she had indeed died of asphyxiation, first by hands and then finished by the underwear. The investigation could not locate any of China Mary's accumulated wealth, and robbery was deemed the motive for the crime. Even the local newspapers understood the lack of effort shown by authorities in trying to find the perpetrator: "There is little likelihood of our energetic and wide awake officials making any great effort to ferret him out" (*Helena Daily Herald* 1867a:3).

The next week the Chinese community performed a traditional burial ceremony with several major exceptions. According to reports the male Chinese community refused to go to her funeral, as she apparently "disowned her country and claimed Japanese nativity, and refused to have anything to do with them, or to conform to their 'company' arrangements" (*Helena Daily Herald* 1867b:3). Only four Chinese women comprised the funeral party. The Chinese women conducted all the major funeral rites, including preparing and transporting the body from her home to the Helena Cemetery. The remainder of China Mary's personal effects and money were sent to her children in Weaverville, California.

In remarkable and exacting detail the *Helena Herald* reporters analyzed the funeral proceedings, providing a glimpse into the reflection of traditional burial rites in Montana Territory. At a location known as a private hospital on West Main in Helena, mourners prepared China Mary's body and soul for the afterworld. The four Chinese women who served as mourners laid out an elaborate buffet of traditional treats at the head of the coffin at the hospital including, "pears, peaches, nuts, candy, sugar, flour and boquets [sic], and at her feet was placed rice, pork steaks, eggs, onions and a glass of brandy" (*Helena Weekly Herald* 1867b:4). They also placed playing cards, dice, and paper money in the coffin, and reportedly $40 of gold coins into the mouth of China Mary. In addition, mourners burned incense and paper money in the hospital room before packing up the coffin for transport to the cemetery (*Helena Weekly Herald* 1867b:4).

At the cemetery mourners threw silver coins, rice, beans, and other foodstuffs into the coffin before it was lowered into the grave shaft. Apparently the reporters believed that the Chinese used silver coins to help preserve the woman's body for disinterment in the future, though this hypothesis is unsubstantiated. The *Helena Herald* staff had an interesting observation; they believed "that the bodies of women are seldom removed, especially those of such women as the one whose funeral we are describing" (*Helena Weekly Herald* 1867b:4). The last rites performed by the Chinese women mourners were to burn all of China Mary's clothing and then to place in her home a final offering of food on a silver platter. Following traditional practices involving their concern for bad spirits, they attempted to convince the landlord of China Mary's home to board it up and not allow anyone to live in it (*Helena Weekly Herald* 1867b:4). Leeson (1885:718) believes that China Mary was the first Chinese individual buried in Helena's cemetery.

The murder of China Mary, while making for good storytelling, touches on several themes within the study of Chinese experiences in Montana. First, the murder was barely investigated by European American authorities, which highlights the low rank Chinese women held in territorial society. It is likely that investigators would have investigated the murder of any male regardless of ethnicity, whereas there was never a suspect in China Mary's murder. Interestingly, in Helena only one month after China Mary's murder, a Chinese man, Ah Weng, attempted to murder by strangulation a prostitute by the name of Chinese Susie. Chinese Susie successfully called for assistance and Ah Weng disappeared, only to be found a few hours later hiding underneath Susie's bed. This time Sheriff Warfield arrested Ah Weng and placed him in jail, though at no time did anyone seem to connect this incident with the murder of China Mary (*Helena Daily Herald* 1867b:3).

Second, China Mary's story illustrates her isolation within the traditional Chinese community of Montana Territory. The lack of male funeral participants and the apparent preclusion of interaction of the Chinese companies and tongs (fraternal societies) suggest the depth of China Mary's social ostracism. Third, China Mary's relationship with the unnamed African American man and the apparent future marriage of the two once reaching California indicates at least some interaction between ethnic groups in 1860s Montana Territory. Finally, the existence of her children in Weaverville, California, provides an example of how Chinese individuals' had familial connections across the United States, information not found in other existing historical documents.

As mentioned earlier, in 1870 and 1880 the federal census recorded a large proportion of Chinese women as prostitutes. Many of the known stories involving the Chinese women of Montana Territory revolved around prostitutes or prostitution. For instance, the first mention of Chinese women in Missoula related to a prostitute who "attempted to build a fire on the floor of her domicile using a piece of tin for a stove" (*Helena Daily Herald* 1870d:3). In another instance, a Chinese woman residing in Helena, reportedly a prostitute, broke out the window of a neighbor's residence, badly lacerating the occupant (*Helena Independent* 1877c:3). Other stories relating to Chinese prostitutes during this early period include the death of Die How in Helena by natural causes and her subsequent burial (*Helena Daily Herald* 1867c:3). It appears that suicide was the most common cause of death in the Chinese prostitute community, and this likely reflects the harsh conditions of their lives. In early 1871 two Chinese women committed suicide in Deer Lodge by consuming large quantities of an unknown poison or opium (*New Northwest* 1871d:3). There was another attempted suicide by a Chinese prostitute in Deer Lodge during 1872, and the newspaper noted that this behavior was not that uncommon among the European American prostitutes as well (*New Northwest* 1872a:3).

In the early 1870s Deer Lodge city commissioners perceived problems with the presence of Chinese prostitutes along Main Street. Even though the 1870 federal census recorded only two Chinese women living in Deer Lodge, a complaint making its way into the territorial grand jury in 1870 stated that an alley perpendicular to Main Street was "inhabited and densely crowded with China women of 'ill repute.' It is charged that the neighborhood is disturbed (many times during the entire night) by the great noise, rowdyism [sic], and firing of pistols" (*Helena Daily Herald* 1870b:3). It is unclear whether many of the Chinese women were missed by the census takers, or if the report was greatly exaggerated, but it seems the former is a more likely scenario.

An incident in 1869 illustrated the Deer Lodge city officials' complaint. Non-Chinese antagonists threw rocks at a Chinese bagnio, or prostitution house, on Main Street. The female Chinese occupant returned the favor with discharge of a pistol, though no one on either side was injured.

The *New Northwest* noted that the Chinese prostitutes on Main Street were a nuisance and even noted that the European American prostitutes of Deer Lodge had at least moved to less busy thoroughfares (*New Northwest* 1869a:3). While the complaint in the Montana territorial court had little effect on the placement of Chinese prostitutes in Deer Lodge, it led to the analysis of failures of the census bureau to record accurately the number of Chinese women in that city. As such the *Helena Daily Herald* (1870b:3) wondered whether "the Grand Jury of Deer Lodge county have not confined themselves to facts in their report, or Deputy Marshal Jones has unaccountably blundered in his census record of Mongolion [sic] female residents of his county." This single example suggests that the numbers of Chinese women in Montana Territory were underestimated, a problem that plagues interpretations of historical Chinese demographics across the United States. A similar omission of Chinese women in the 1870 census records was found in other areas of the American West including Virginia City, Nevada (James 1998:96).

With so few Chinese women in Montana Territory, stories of weddings were rare, yet heavily covered in contemporary newspapers. The first such Chinese wedding likely recorded in Montana Territory occurred in 1870 in Deer Lodge. In early 1870 Wah Lee purchased the rights to a Chinese woman for $600 and imported her to Deer Lodge. Jo, a Chinese laundry employee, sought to marry the woman but needed to raise the resale price of $950. Jo was unable to acquire the required amount, and Wah Lee started the woman on a trip to the Cedar Creek mines, likely for use as a forced prostitute. Upon hearing from a European American that all he needed to do was to marry her according to American law, Jo set out after the woman and Wah Lee. Jo overtook Wah Lee with the assistance of a law officer and married the woman without paying the agreed-upon price (*New Northwest* 1870a:3). In 1871, the year after the initial excitement, the Chinese wife of Jo gave birth to a son, the first recorded Chinese birth in Deer Lodge County and perhaps Montana Territory (*New Northwest* 1871b:3).

Sometime during the next two years another child arrived for Jo and his wife. In 1873 a Chinese suitor approached Jo's wife, and they disappeared together to nearby Pioneer. After Jo discovered that his wife left with $1,000 and her wedding gifts, he apparently contacted the local Chinese community leader and dispatched a call for their arrest. Upon arriving in Pioneer, Jo found his wife full of opium and whisky with the Chinese suitor disappearing with all the money and gifts. In her intoxicated condition she fought against Jo and his efforts to return her to Deer Lodge, but eventually both made it back to their town (*New Northwest* 1873c:3). After returning to Deer Lodge, the couple appear to have settled their differences, and they disappeared from the historical record.

Throughout the 1870s there were other instances of violence and legal fights over Chinese women in Montana Territory. In 1872 a Chinese man robbed a fellow countryman in Pioneer of $750 in cash and a woman and headed for Deer Lodge. After eluding the owner of the Chinese woman for several days, the pair finally managed to sneak out of Deer Lodge and leave the territory (*New Northwest* 1872i:3). A month later in Philipsburg, two Chinese men gambled over the ownership of a woman, which led to a deadly resolution. After several hours of gambling the first Chinese man, who lived in Philipsburg, put up the woman as part of the bet and lost her. The second individual, a Chinese man of Pioneer, took the woman back with him to that settlement as part of his winnings. A few days later the Chinese man from Philipsburg went to Pioneer, used a knife to murder the woman, and then disappeared onto a stage heading out of the territory (*New Northwest* 1872j:3). It is unclear if the Chinese man from Pioneer is the same in both these stories, though it is possible.

In yet another story of the apparent commodity role Chinese women played in early Montana Territory, a Chinese restaurateur used his wife as the collateral for a loan. Charlie, a Chinese man living in Yreka, used his wife of five years as a $1,000 security deposit for purchasing a large stock of goods from Quong Lee of Pioneer. After a slow year on the Elk Creek mines, Charlie was unable to pay the debts owed to Quong Lee and the wife was transferred. After changing hands, Charlie's wife pleaded with European American miners in Pioneer to secure her release, as her new role was likely as a prostitute in the camps. At last report in 1874 the miners were in negotiations with Quong Lee to secure the Chinese woman's release and employ her as a domestic servant with a family in Deer Lodge (*New Northwest* 1874c:3).

By the late 1870s Gem Kee, who owned and operated the large Kem Lung Chung store in Deer Lodge, was wealthy enough to purchase a wife from China. Due to discrepancies in the historical record, and a variably changing name, this store might also be the earlier described Kung Chung Lung Company or the Kim Chung Lung store. The 1880 federal census listed Gem Kee (Gimgee in the rolls), as a single 32-year-old merchant living with several other Chinese men, including his married business partner Buck Gim (Federal Census Bureau 1880). In late fall of 1880, after the census enumeration, Gem Kee expanded the Kem Chung Lung building in Deer Lodge with a second floor solely for the purpose of housing his new wife. Gem Kee left for San Francisco in October 1880 to marry a "Little Footed" or "bound foot" Chinese woman. The *New Northwest* noted that Gem's wife might have been the first bound foot woman brought to Montana Territory (*New Northwest* 1880c:3), and they arrived in Deer Lodge in November 1880 (*New Northwest* 1880d:3).

In late November Gem Kee sent out 400 invitations to the European American and Chinese population of Deer Lodge and vicinity to meet his new wife. J. H. Oliver, the same European American individual who translated for the Chinese witnesses in the Martell manslaughter trial in Blackfoot City in 1871, served as the party's interpreter. According to the *New Northwest* reporter, who was also invited, more than 100 Chinese residents of Deer Lodge and vicinity came to the party with gifts including gold watches and bracelets. Apparently when Gem Kee arrived in San Francisco to marry the betrothed woman, she refused to wed due to the harsh climate of Montana Territory and the lengthy travel. Gem Kee found a replacement and married a 20-year-old Chinese woman by the name of Ida. According to Oliver, Ida was a member of a wealthy and elite family in Canton Province. As was the fascination with all European American guests, they measured Ida's foot out at 3½ inches long by 1¼ inches wide at the instep. Because of Ida's limited mobility, Gem Kee employed several servants to attend her and assist with her chores and walking (*New Northwest* 1880d:3). In December 1880 Gem Kee's business partner, Buck Gim, headed to California to acquire a similar bound foot wife (*New Northwest* 1880e:3), though the census record already noted that he was married.

There is far more to the lives of Chinese women in early Montana Territory than prostitution and fights over ownership and marriage, though due to the biased nature of the European American majority and Chinese men, those stories are lost. A marriage between a European American man and Chinese woman in Blackfoot City illustrates the true depths of racial bias. In 1873 Benjamin Harris married Ah Fu in Blackfoot City, while using J. H. Oliver as translator and marriage officer. The marriage led the *New Northwest* to question why an "intelligent, respectable, American citizen" would marry "one of the miserable creatures known as China-women, the lowest, most degraded of the race, precluded maternity and dedicated to shame" (*New Northwest* 1873g:3). In addition, the reporters suggested that "the ceremony . . . is therefore illegal and void . . . and it would be [a] better course . . . to renounce the unholy, unnatural alliance, and leave the pagan to her kind" (*New Northwest* 1873g:3). It is apparent that if positive stories about Chinese men were rare in Montana Territory, an unbiased position toward women was nearly impossible to find. Thus, when discussing the Chinese women of Montana Territory, it is important to remember that few of the positive aspects of their experiences made it into the historical record. Moreover, there appear to be no sources providing a female perspective—whether positive or negative—on life in early Montana Territory.

## Population Growth but Further Legal Storms on the Horizon

During the nearly two decades after Montana became a territory, the Chinese population spread throughout all the major and minor mining districts and townships. Having a Chinese population within a town or settlement boundary was by far the rule and not the exception. Throughout the 1860s and 1870s Chinese immigrants moved into placer gold districts across the territory, seeking their fortune in large and small groups. In many districts the Chinese population was significant enough to support several stores dedicated to the importation of their goods direct from China. It is unclear how much gold in Montana Territory Chinese miners recovered, but guarded estimates would place it in the $8–$15 million range. It is clear that the early Chinese population met severe social and legal discrimination, yet overcame all challenges to create a vibrant community supporting traditional values. The initial period of the Chinese in Montana is marked by unsuccessful attempts to exclude the population legally from the mining and laundry enterprises. As the century progressed, growing anti-Chinese sentiment across the United States was about to boil over and begin the systematic exclusion of this population. In this new period the anti-Chinese policies of the 1860s and 1870s would find new vitality and would have dire consequences for the immigrants fighting to stay atop *Gum San*, Gold Mountain.

## 2. Restriction and Legal Attacks, 1880–1900

### The 1880s–1900

For the Chinese in Montana, the period between 1880 and 1900 included some of the most significant accomplishments and some of the vilest acts of racial depredation against a single ethnic group in U.S. history. In the 1860s and 1870s the Chinese, while persecuted, had a level of legal protection governed by the rule of law that often went in their favor, such as the Montana Territory's Alien Law that passed in 1872 but struck down by 1874. By the end of the economic recession that afflicted the United States in the 1870s, the climate of the legal system changed, and the passage of Montana's Opium Law in 1880 and the federal Chinese Exclusion Act in 1882 helped lead to the eventual collapse of the state's Chinese population. By the end of the 19th century the best years experienced by the Chinese population of Montana had ended, and the future held little financial or social promise.

By the 1880s the Chinese in Montana Territory had established significant centers of trade and commerce in central settlements like Helena, Deer Lodge, Butte, and Fort Benton. In the outlying mining districts the Chinese population became established and created vibrant, mostly male communities by successfully purchasing and working previously worked claims. The boom years of Montana's placer camps had passed, and the Chinese now working these claims were on borrowed time in terms of financially prosperous mining endeavors. As demonstrated in the 1880 federal census of Montana Territory discussed later, there was a decided shift occurring from mineral exploration to service industries such as laundries and restaurants.

Historical documents from the 1880–1900 period of Montana Chinese history provide more substantial information than in the previous period discussed. Overall, there was increased visibility of the Chinese of Montana in the historical records during the last two decades of the 19th century, but much of the material still focuses on mundane facts of leases, deeds

Table 6. Chinese archaeological sites in Montana relating to the 1880–1900 period

**Confirmed Chinese Archaeological Sites**

| Site # | Site Name | Period | Owner* |
|---|---|---|---|
| 24BE0169 | Bannack | 1862–1880S | State of MT |
| 24BE0187 | Glendale | 1870–1900 | Private/BLM |
| 24BE0807 | Lion City | 1870–1880S | University of MS |
| 24BE1381 | French Creek Cabins | 1870–1880S | USFS |
| 24BE2186** | Blue-Eyed Annie | 1895–1901 | USFS |
| 24BW0497 | Lombard | 1890–1910 | BLM |
| 24GN0363 | Granite | 1880–1893 | USFS/Private |
| 24GN0377 | Sierra Mine Site | 1870S–1910S | BLM/Private |
| 24GN0540 | Garnet | 1880S–1930S | BLM/Private |
| 24LC0712 | Chinaman's Cove | 1860S–1880S | BOR |
| 24LC1083 | Marysville | 1870S–1900S | Private |
| 24LC1711 | Upper Cave Town | 1868–1880S | USFS |
| 24LC1723 | Upper Cave Gulch | 1870S–1890S | USFS |
| 24MA0723 | Virginia City | 1863–1890S | Private/State |
| 24MN0249 | Louiseville | 1870–1890S | USFS |
| 24MN0332*** | Rowley Chinese Hearths | 1870–1890S | USFS |
| 24MO0110 | Montreal or "Old Town" | 1874–1890S | USFS |
| 24MO0172 | Coloma | 1880S–1910S | BLM/Private |
| 24MO0266 | Fort Missoula | 1877–PRESENT | County/USFS/Private |
| 24PA1330*** | Huckleberry Camp | 1897–1900 | USFS |
| 24PW1021 | Blackfoot City | 1865–1880S | Private/BLM/USFS |
| 24SA0155 | Weeksville | 1881–1883 | Private |
| 24SA0591*** | West Noxon Chinese RR Camp | 1881–1883 | Private(MT Rail Link) |
| 24SA0592*** | Submerged Noxon Chinese RR Camp | 1881–1883 | Private (Avista Corp.) |
| 24SA0593*** | Thompson River Chinese RR Camp | 1881–1883 | Private(MT Rail Link) |
| 24SA0594*** | East Eddy Chinese RR Camp | 1881–1883 | Private(MT Rail Link) |
| 24SA0595*** | East Weeksville Chinese RR Camp | 1881–1883 | Private(MT Rail Link) |
| 24SA0596*** | "Last Chance" Railroad Camp | 1881–1883 | Private(MT Rail Link) |
| 24SA0597*** | "Fu Sang" Chinese RR Camp | 1881–1883 | Private(MT Rail Link) |
| 24SA0598*** | "Taklamakan" Chinese RR Camp | 1881–1883 | Private(MT Rail Link) |
| 24SA0599*** | Perma Bridge Railroad Camp | 1881–1883 | Private(MT Rail Link) |
| 24SB0212 | German Gulch | 1865–1900 | USFS/Private |
| 24SW0738 | Big Timber | 1882–1940 | Private |

Table 6. Chinese archaeological sites in Montana relating to the 1880–1900 period

**Potential or Unconfirmed Chinese Archaeological Sites**

| | | | |
|---|---|---|---|
| 24BE0806 | Hecla | 1870s–1880s | University of MS |
| 24BE2003 | Cow Creek Cabins | 1870s–1880s | USFS |
| 24GN0894 | Bedrock Flume or "China Diggings" | 1860s–1890s | Private/BLM |
| 24JF0290 | Chinese Diggings | 1870s–1880s | Private |
| 24LC0467 | Old Lincoln | 1865–1880s | USFS |
| 24LC1112 | Austin Mining District | 1870s–1880s | USFS/Private |
| 24LC1716 | Upper Cave Town | 1870s–1880s | USFS |
| 24MO0179 | Yreka | 1860s–1890s | State/UM/Private |
| 24PW0183 | Reynold's City | 1860s–1890s | BLM/Private |
| 24SA0105 | China Ditch | 1885 | USFS |
| 24SA0348 | NPRR Eddy/Weeksville | 1881–1883 | Private |
| 24SA0568 | White Pine | 1881–1910 | Private |

* BLM = Bureau of Land Management, USFS = United States Forest Service, BOR = Bureau of Reclamation.

** = New site in 2008, no Chinese artifacts.

*** = New site discovered in summer of 2008.

and mineral claims, and anecdotal stories in newspapers. Archaeological studies of sites within this second period of Montana Chinese history provide an additional source of evidence about the lives of these pioneers. There are currently 33 confirmed archaeological sites recorded in Montana that are associated with the Chinese during this period. Fifteen of these 33 sites are continuations from the first period, as the arbitrary temporal boundaries used here do not neatly fit the occupational history of some areas. For instance, most of the Chinese occupation and mining in German Gulch occurred in the 1870s, but populations continued to work and live in the area throughout the 1880s and 1890s. In addition, there are 12 other sites that might provide archaeological association with the Chinese but are as yet unproven or undocumented (Table 6).

## Demographics in 1880

The 1880 Montana census enumerated 1,763 Chinese in 10 counties (Figure 7 and Table 7). Swartout (1988:48) notes that there were 1,765 Chinese enumerated in the 1880 census, but careful transcription discovered that two individuals were actually listed twice in separate areas of the same county, making the correct total 1,763. From 1870 to 1880 the recorded Chinese population dropped by 184, most likely the direct result of the recession that

Table 7. Montana Chinese population by county, 1880 Federal Census

| County | Chinese Population | | |
|---|---|---|---|
| | Male | Female | Total |
| Beaverhead | 124 | 6 | 130 |
| Choteau | 18 | 0 | 18 |
| Custer | 2 | 0 | 2 |
| Deer Lodge | 677 | 31 | 718 |
| Gallatin | 27 | 1 | 28 |
| Jefferson | 52 | 0 | 52 |
| Lewis and Clark | 329 | 30 | 359 |
| Madison | 258 | 7 | 265 |
| Meagher | 50 | 2 | 52 |
| Missoula | 144 | 5 | 149 |
| Totals | 611,681 | 82 | 1,763 |

hit the United States in the 1870s and a shift away from placer mining. Lewis and Clark County has the most substantial decrease in Chinese population, dropping by more than 50% during the preceding decade. This was a result of Helena and surrounding mining districts passing their peak production and shifting toward heavily capitalized lode mining. Montana counties with substantial mining endeavors still possessed the largest Chinese populations, including Deer Lodge (718), Lewis and Clark (359), Madison (265), and Beaverhead (130). In addition, Missoula County, with Cedar Creek mining operations continuing, increased Chinese population from 44 in 1870 to 149 in 1880. Similar to 1870, Deer Lodge County was the largest center of Chinese population in Montana. Before 1881 Deer Lodge County incorporated a significant swath of southwest Montana's mining communities, including German Gulch, Butte, Silver Bow, Rocker, Blackfoot City, and the Pioneer and Gold Creek workings, which explains the high number of Chinese residents. Due to the decline of placer mining opportunities in Helena and Virginia City vicinities, Deer Lodge County's Chinese population went from 18% of the state's total in 1870 to over 40%. Even with Missoula County's substantial increase, nearly 89% of the Chinese population was still centered in the mining districts of southwest Montana, though employment was shifting away from mineral exploitation.

In individual locations (Table 8) Butte, Alder Gulch and Virginia City, and Pioneer and Cedar Creek dominated the population numbers, all having over 100 Chinese residents, with the highest in Helena at 287. Smaller but significant populations of Chinese existed in other distinct settlements, like Deer Lodge and Bozeman, and in areas without formal townships,

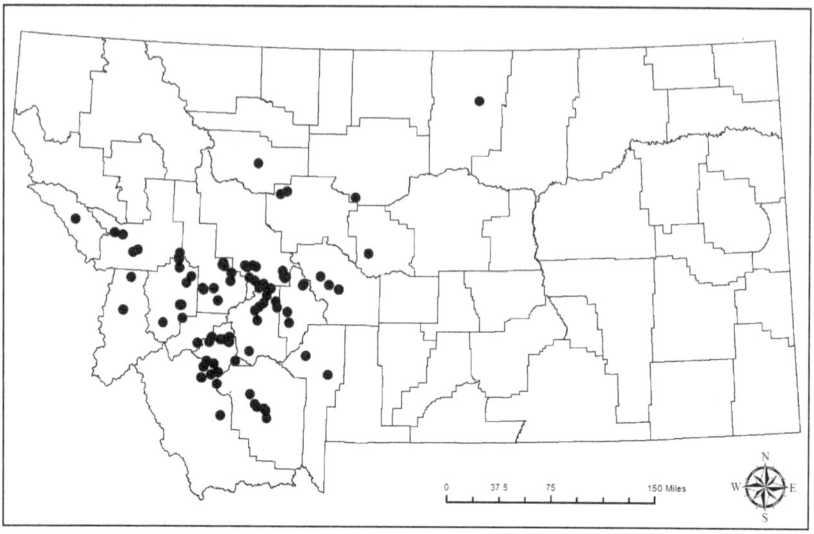


Figure 7: Chinese population centers in 1880. (Created by author.)

like German and French gulches or Elk and Silver Bow creeks. As evidenced by both German Gulch and Cedar Creek settlements discussed in the previous chapter, the dispersed settlements in 1880 probably had a relatively high degree of autonomy and independence within the larger European American population of Montana Territory. More significantly for understanding the Montana Chinese experience, there was a proliferation of small populations moving into more dispersed settlements where there were few or no placer mining opportunities. In 1870 there were 37 settlements or enumeration districts with Chinese populations or individuals, but by 1880 there were 88. In rural communities like New Chicago, White Sulphur Springs, and Hamilton, the economy focused more on commercial or agricultural interests, and Chinese pioneers seized on the opportunity to establish laundry and restaurant businesses.

The 1880 census provides the first indication of some major shifts in Chinese demographics in Montana Territory, including age (Figure 8) and employment. In 1870 the average age of a Chinese immigrant, regardless of sex, was 30, but by 1880 it had increased to 35. The average ages of Chinese men and women increased from 1870 to 1880 as well. In 1870 the average age of men was 30 while women were a bit lower at 25. In 1880 the average ages increased to 35.4 and 28.4 for men and women respectively. A married couple living in Virginia City named

Table 8. Montana Chinese population by location, 1880 Federal Census

| Location | # Chinese | Location | # Chinese |
|---|---|---|---|
| Helena | 287 | Silver Bow | 4 |
| Butte City | 178 | Confederate Gulch | 4 |
| Alder Gulch | 166 | Ten Mile/Greenhorn | 4 |
| Pioneer City | 123 | Bearmouth | 4 |
| Cedar Creek | 117 | Hamilton | 4 |
| Virginia City | 91 | Bitterroot Valley | 4 |
| Beaverhead River | 90 | Frenchtown | 4 |
| Deer Lodge | 50 | Diamond City | 3 |
| Ophir Gulch | 45 | Park City | 3 |
| Elk Creek | 44 | Vestal | 3 |
| German Gulch | 39 | Fort Logan | 3 |
| French Gulch | 38 | Clancy | 3 |
| German & Silver Bow | 37 | Ft. Missoula | 3 |
| Flint Creek Valley | 32 | Dewey's Flat | 2 |
| Bozeman | 25 | Timber | 2 |
| Glendale | 24 | New Chicago | 2 |
| Yam Hill | 21 | New York Gulch | 2 |
| Prickley Pear [sic] | 21 | Yogo Gulch | 2 |
| Georgetown | 18 | White Sulphur Springs | 2 |
| Belmont | 18 | Gloster | 2 |
| Washington Gulch | 17 | Smith River Valley | 2 |
| Fort Benton | 16 | Mitchell Gulch | 2 |
| Missoula | 16 | West Gallatin Valley | 2 |
| Hasmarck | 15 | Trapper Gulch | 1 |
| Cave Gulch | 15 | Quartz Hill | 1 |
| Snowshoe Gulch | 14 | Big Hole | 1 |
| Boulder | 13 | Upper Teton River | 1 |
| Lion's City | 12 | Fort Belknap | 1 |
| Oregon Gulch | 10 | Miles City | 1 |
| Cooper Gulch | 9 | Junction City | 1 |
| Beartown | 9 | Laurin | 1 |
| Saint Louis | 8 | Sheridan | 1 |
| Grizzly Gulch | 8 | Adobetown | 1 |
| Philipsburg | 7 | Silver City | 1 |
| Walkerville | 6 | Mount Pleasant | 1 |
| Wickes | 6 | Radersburg | 1 |
| Ruby Valley | 6 | Crow Creek Valley | 1 |
| Rocker | 5 | Blue Cloud | 1 |
| Sun River | 5 | Soap Gulch | 1 |
| Fort Shaw | 5 | Upper Highland | 1 |
| Pipestone | 5 | Comet Mines | 1 |
| Jefferson City | 5 | Hellgate/Little Blackfoot | 1 |
| Dry Gulch | 5 | Browne's Gulch | 1 |
| Trinity Creek | 5 | Frenchtown & Six Mile | 1 |

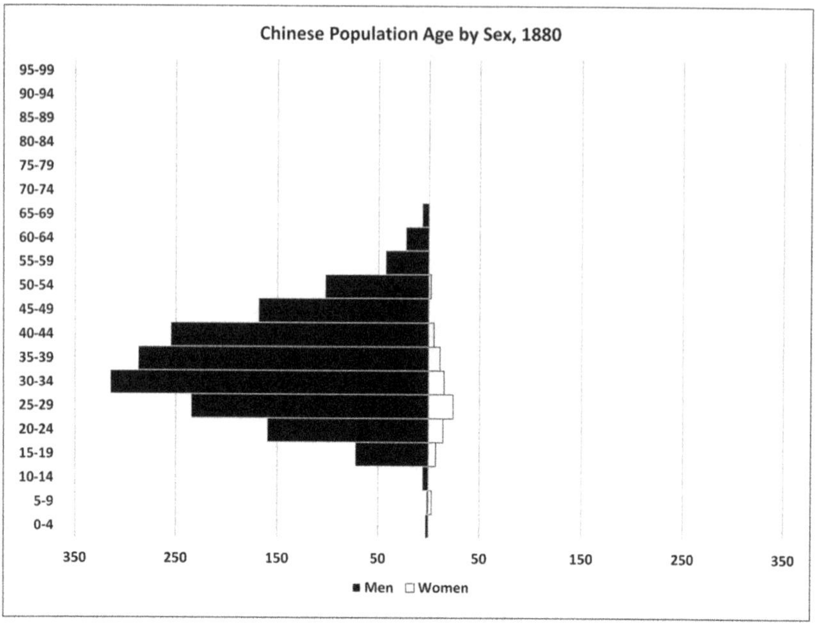


Figure 8: Population pyramid for Chinese in Montana, 1880. (Created by author.)

La Won and Sing Choy had the two youngest Chinese children in the territory. Both children were born in Montana Territory and were aged 1 for the boy, and 8 for the girl. An 83-year-old placer miner living in Cedar Creek was the oldest man, while two 50-year-old females were the oldest women.

In 1870, 83% of all Chinese living in Montana Territory were engaged in mining operations. Over the course of the next 10 years the number of placer booms dwindled, and the economy shifted toward more stable lode mining or into the agricultural and ranching sectors. Because of this economic transition, the Chinese population also began to shift toward a more service-based orientation. Employment associated with mining still accounted for 59.6% of male occupations, but opportunities in cooking and laundry situations accounted for the most substantial growth for the Chinese in Montana Territory. Employment as cooks in private homes, sawmills, mines, hotels, or restaurants accounted for 14% of Chinese male employment, an increase from only 2% of the workforce in 1870. Chinese working in the laundry business accounted for 10% of occupa-

Table 9. Montana Territory employment of Chinese men, 1880 Federal Census

| Occupation | # of Chinese Employed |
|---|---|
| Mining | 1,002 |
| Cook* | 237 |
| Laundry | 174 |
| Unknown | 42 |
| Gambler | 38 |
| Laborer | 31 |
| Loafer | 30 |
| Servant | 25 |
| Merchant/Store Keeper | 22 |
| Gardener | 20 |
| Wood Chopper | 11 |
| Barber | 7 |
| Restaurant | 6 |
| Butcher | 4 |
| Dishwasher | 4 |
| Doctor | 3 |
| Prisoner | 3 |
| Baker | 2 |
| Grocer | 2 |
| Priest/Clergy | 2 |
| Teamster | 2 |
| Acrobat | 1 |
| At Home | 1 |
| At School | 1 |
| Blacksmith | 1 |
| Bookkeeper | 1 |
| Cheese Maker | 1 |
| Chore Boy | 1 |
| Druggist | 1 |
| Engineer | 1 |
| Gambling House Keeper | 1 |
| Stock Raiser** | 1 |
| Store Clerk | 1 |
| Tailor | 1 |
| Waiter | 1 |

* George Taylor, cook, was recorded twice in the 1880 census.

**Tommy Haw, a stock raiser, was recorded twice in the 1880 census.

tions, up from 6% ten years prior. The increase of employment as cooks, in laundries, or as servants from 1870 to 1880 foreshadows trends that continued during the late 1800s and early 1900s.

Perhaps a result of more accurate census enumeration, there appears to be a diversification of more specialized labor categories (Table 9). Some notable Chinese jobs listed in the 1880 census include butcher, barber, cheesemaker, druggist, tailor, acrobat, and stock raiser. As in 1870, Helena had two Chinese doctors in 1880, but they appear to be different individuals. Most notably, the acrobat and stock raiser stand out as unique categories. The acrobat was listed as Ung Gay, a 20-year-old man, living in Helena. It is likely that Ung Gay was a member of a traveling theater group operated by non-Chinese, but this is hard to substantiate. Tommy Haw was actually recorded twice in Beaverhead County as a stock raiser, probably a result of his moving to various properties he owned in the area. Haw and his interesting life history are discussed in more detail later in this chapter. Similar to Tommy Haw, enumerators listed George Taylor twice in the 1880 census, both times as a cook.

Between 1870 and 1880 Montana Territory's Chinese female population declined from 125 to 82. Swartout (1988:48) states that there were only 80 Chinese women Montana Territory at this time, and this discrepancy is likely from miscalculation by the census bureau. Most of the drop in women's numbers occurred in the large Chinese population centers in Deer Lodge and Lewis and Clark counties. As noted in chapter 1 the sex ratio in 1870 was a significantly high 14.23 males per female, but by 1880 the situation worsened with a dramatic rise to a ratio of 20.51 males per female. In contrast, the total proportion of men to women in Montana Territory, regardless of ethnicity, was 4.2 in 1870 and 2.5 in 1880. Chinese women had fewer employment opportunities than their male counterparts in 1880 (Table 10). In 1870 prostitution and a vague category of "Keeping House" were the two largest employment categories for Chinese women. This pattern continued in 1880, though prostitution increased its proportion of the employment category from 44.5% in 1870 to 74.4%. Problems with how census enumerators determined women's role as prostitutes were noted in chapter 1. "Keeping House" declined from 41.4% of Chinese female employment to only 9.8%. Most of the Chinese women engaged in "Keeping House" were married or widowed.

Table 10. Montana Territory employment of Chinese women, 1880 Federal Census

| Occupation | # of Chinese Employed |
|---|---|
| Prostitute | 61 |
| Keeping House | 8 |
| Unknown | 8 |
| House Keeper | 2 |
| Washing | 2 |
| Servant | 1 |

## Montana's Opium Law

As population centers of Chinese immigrants spread across Montana Territory, the use and abuse of opium followed them. Across the United States, thousands of Chinese and non-Chinese became addicted to opium, in either its smoking or medicinal (liquid) form. Opium smoking was as ubiquitous within the Chinese community as alcoholism was in the neighboring European American households. Even newspaper accounts of the period indicate the similarity between alcohol abuse in European American populations and the Chinese use of opium (*Helena Independent* 1880:3). It is highly likely that the first Chinese immigrants who came to Montana Territory following the gold bonanzas of the early 1860s brought with them the tools and supplies for the smoking of opium. The reasons for the use of opium by the Overseas Chinese populations are a well-trodden area of historical and archaeological research (Ritchie 1986; Wegars 1993; Ahmad 2007). Recently some Chinese American scholars have attacked the historical and archaeological focus on opium smoking as merely perpetuating long-standing racial stereotypes and prejudices against these immigrants (Fong 2007).

Regardless of modern political correctness (e.g., Fong 2007), opium smoking was a common recreational and social activity in the rural and urban populations of Montana Territory but not only by the Chinese (see also Felton et al. 1984). Some estimates state that only 20% of the opium imported into the United States during the 19th century was actually used by the Chinese, with the remaining 80% being used in popular European American medicinal concoctions such as laudanum (Wylie and Fike 1993:258). The use and abuse of the medicinal variety, opium mixed with alcohol, was pervasive in 19th- and early 20th-century white middle-class America, primarily in the form of laudanum (Goldberg 2006:7–8). Consumers even used an opium derivative to soothe the aching gums of teething babies (Goldberg 2006:6). Many European Americans, from both the working and upper classes, also smoked opium recreationally and even became addicted (Wylie and Fike 1993:258).

Regardless of the user, the smoking of opium generally occurred within a Chinese establishment. It appears that there were rarely entire establishments in Montana Territory devoted to the use of opium; rather, people often smoked opium in a communal setting within a building used as a laundry, restaurant, saloon, or gambling house. Archaeological work at a Chinese laundry and restaurant in Big Timber, Montana, discovered a density of opium-related artifacts in a small addition to the rear of the structure (Moschelle 2009). In addition, a police report indicates there was a raid on another Chinese restaurant in Big Timber due to its dual role as an opium smoking establishment (Moschelle 2009:24–25). While historical maps and records do not indicate the use of the building as an opium smoking establishment, the density of materials suggest that at least this small addition served that function. In another instance, recollections by Davis (1967:2) of the Virginia City Chinese suggest that at least in that community they smoked opium on the ground floor of the Chinese Temple located on Main Street.

A detailed newspaper account from 1880 on Chinese opium smoking in Helena provides description of the physical layout and accoutrements of a "den," and is similar to accounts from other areas (Ritchie 1986; Wylie and Fike 1993).

*The door admits you to a main front room, generally lighted by two windows, one at either side of the entrance. There is generally no other windows in the buildings, except sometimes there is one in the rear of the building. From the front room running clear to the rear is a hall or passageway lined on either side by small rooms not unlike staterooms on steamers—except in the meagerness and plainness of their appointments—and each containing two or four bunks like the berths in an ocean vessel. On the bottom of each berth is a clean matting and along the back runs a common pillow. Rather an uninviting couch for a comfortable snooze, but the Chinamen pass most of their time upon these pallets, and by the magic aid of their opium pipes, not only forget their present cares, but are transported to scenes of transcendent loveliness, which would not make them exchange their humble lot for the throne of a king or the crown of an emperor. The lamp over which the pipes are lighted stands in the middle of the berth and the smoker curls up at one end. (Helena Independent 1880:3)*

This description of an opium den in Helena provides some context to this practice in Montana and is similar to an account by Davis (1967:2) of a den in Virginia City, Montana.

The editorial continues by noting how deleterious the smoking of opium is to human health, discussing both the physical and mental toll of habitual use. Finally, the editorial provides some context for the financial cost of the drug habit, noting that opium purchased from a Chinese store in Helena cost approximately $15 to $17 per can, an increase of several dollars over the preceding year. Wylie and Fike (1993:262) note that in the 1870s the cost for a five-ounce can of opium on the West Coast was $5–$9, and by the early 1880s the price was up to $10 in Tucson, Arizona. The 1880 prices in Montana were higher than in either locale, likely a result of long-distance shipment and the lack of a main transportation network like an east-west railroad. To put these prices into context, the average Chinese placer miner brought in only several dollars a week during the 1870s, making the purchase of opium a substantial drain on financial resources. Today cost of a can of opium indexed to 1880 Helena prices would be $325–$370 when accounting for inflation (Williamson 2010).

While there was opium smoking in Montana throughout the 1860s and 1870s, as evidenced by some newspaper accounts and the archaeological assemblage, a wave of anti-opium agitation and legislation occurred in the territory in 1880. Before 1880 there were only a handful of newspaper stories discussing the use of opium within the Chinese and surrounding communities, and most merely described the use of opium. Opium use seemed almost to appear out of nowhere in the newspaper accounts of the early 1880s, as if Montana citizens were faced with a sudden and previously unknown epidemic. Some Montanans even substituted the restriction of opium as a substitute for the inability to force the Chinese to leave the territory: "If the Chinese can't be made to go, the opium must" (*Helena Independent* 1882a:3).

An 1880 newspaper article in Deer Lodge illustrates the apparent novelty of opium use in Montana. The article listed all the modern conveniences that Deer Lodge lacked, including "no steam fire engine; no 'Great Expectations'; no mineral springs; no 'boom,'" but they did have an "opium den" (*New Northwest* 1880a:3). According to the report the den was located a few doors down from the Chinese store in Deer Lodge, likely the Kung Chung Lung Company store operated by Gem Kee. While the presence of an opium den in Deer Lodge was remarkable to the newspaper editors of the time, it is likely that the establishment, or one similar, had operated in the town since the late 1860s without much notice. So what had changed? The newspaper provides an answer: "Some of the less reputable persons . . . and even white women are alleged to have haunted its shades and reveled in the fumes of the noxious drug until midnight passed" (*New Northwest* 1880a:3). According to a Mr. Spooner, a night watchman for Deer Lodge, on a single night fourteen individuals used the den, and they came from mixed age, sex, and ethnic backgrounds.

Deer Lodge was not the only community to have non-Chinese individuals partaking in the smoking of opium; instead, it was quite a common occurrence throughout Montana in the 1870s and 1880s and into the 1900s. In the fall of 1880 "some of the white women of questionable character are beginning to practice the evil habit, and other respectable white people occasionally visit the dens and are in a fair way to become slaves to the terrible habit" (*Helena Independent* 1880:3). Under Butte's city ordinance, police raided a suspected opium den in Chinatown and arrested several of its occupants, including John Norton, a well-respected member of a minstrel and acting troupe out of Helena and Butte (*New Northwest* 1880b:3).

In another instance John Dunn, a resident of Jefferson City south of Helena, committed suicide by means of overdosing on opium. In this case Dunn was denied morphine from a store and in response went to a nearby Chinese washhouse, where he smoked two pipes of opium before dying in the street outside the establishment (*New Northwest* 1881c:3). According to a subsequent coroner's inquest, Dunn was most likely placed outside the washhouse by the proprietors in a sitting position, and in his opium-induced coma his head dropped, cutting off the flow of air, effectively causing death by self-strangulation (*Helena Independent* 1881g:3).

Perhaps as an impetus for the Montana Opium Law proposed several months later, the *New Northwest* editors suggest that the territorial legislature pass an act that could protect unincorporated settlements from the vice for the common good of the citizenry's health (*New Northwest* 1880a:3). Proponents of new legislation claimed it to be for the protection of the non-Chinese citizens of Montana Territory from the "horrible and revolting vice of opium eating and smoking," as it was "spreading and rapidly undermining the health and happiness of the young and inexperienced of both sexes" (*Helena Independent* 1883b:3). Ahmad (2007:40) notes that between 1881 and 1900 the Warm Springs Asylum, outside Deer Lodge, Montana, housed eight Chinese men who were likely committed for opium-related mental ailments.

It is unclear exactly what precipitated the passage of Montana's Opium Law in the winter of 1880–1881, but it was likely a diffusion of similar legislation from other states and territories in the American West around that time. Nevada passed an anti-opium law in 1877, and by 1900 several other western states had followed suit (Erlen and Spillane 2004:11). It appears that local ordinances were also the first vestiges of the upcoming territorial opium law in 1880. For instance, Helena and Butte's city councils sponsored ordinances making the operation of smoking places and distribution of opium illegal in the incorporated city limits in 1879 (Ahmad 2007:58). Attempts by the Chinese to overturn the local ordinances failed, as demonstrated by the case of China Joe. He was arrested for opium smoking in 1880, acquired European American legal counsel to fight the local anti-opium ordinance, but lost the case and was still fined $50 (Ahmad 2007:65).

On February 22, 1881, the Montana Territorial Legislature passed the Opium Act as a means of suppressing the operation of opium dens (Ahmad 2007:59), and it became effective on the first day of May of that same year (*New Northwest* 1881a:3). Several territorial newspapers including the *New Northwest* carried the full text of the Opium Act. Montana's Opium Act did not specifically make opium an illegal drug; rather the provisions focused on closing opium smoking establishments. Section 1 of the Opium Act made it illegal to use a private property "for the purpose of indulging in the use of opium, or any preparation containing opium, by smoking or otherwise, or who shall sell opium for such purpose" (*New Northwest* 1881a:3). Sections 2 through 5 of the Opium Act provided the punitive repercussions for being caught in the smoking of opium, including liens being applied to owned property and forfeiture of leases. Finally, Section 6 set the maximum fine for conviction under the Opium Act at $500 and/or imprisonment not to exceed six months (*New Northwest* 1881a:3).

After the Opium Act came into force on May 1, 1881, there were numerous stories detailing the accounts of raids on opium dens or joints by officers in various Montana communities. Before passage of the Opium Act, there were but a few raids on opium dens under the auspices of local ordinances, as discussed earlier. In the next few years local law enforcement officers arrested dozens of Chinese and whites in places such as Butte, Helena, Fort Benton, and Missoula (*Helena Independent* 1881j:3, 1883c:3; *Weekly Missoulian* 1882a:3).

It would be necessary to analyze the criminal records of these settlements to determine the exact number of individuals prosecuted under the new Opium Act, though that was outside the scope of this project. A cursory survey of the *Helena Independent* between 1881 and 1884 discovered 27 confirmed arrests in Butte, Helena, and Livingston alone. Some opium smokers were released without going to trial based on lack of evidence, which also affects the proper counts (*Helena Independent* 1881j:3). The single largest bust of opium smokers under the Opium Act occurred in the railroad boomtown of Livingston, where police arrested 10 Chinese and European Americans (*Helena Independent* 1883c:3). An April 1882 raid on an opium den in Missoula captured nine Chinese men including the operator of the establishment (*Weekly Missoulian* 1882a:3).

Perhaps no other locale of similarly sized population prosecuted the opium law with such apparent vigor as Fort Benton (see Robison 2009b), located on the Missouri River. With its share of rough and rowdy individuals, many non-Chinese shared in the practice of opium smoking. During Chinese New Year festivities in Fort Benton in February 1880 it appears that some European Americans partook in the activities. For example, the *Benton Record* noted, "How many opium dens those ladies graced with their presence, I am unable to say, but presume that their curiosity was satisfied to the fullest extent" (*Benton Record* 1880:2). Even after passage of the Opium Act, editorials ran in local newspapers noting that European Americans of rough and genteel natures shared in the practices of opium smoking in Fort Benton's Chinese quarter (*Fort Benton River Press* 1881b:2).

Beginning in January 1881, a month before the Opium Act even passed the territorial legislature, there were reports that the sheriff in Fort Benton was already watching Chinese businesses and homes for clues to opium use (*Fort Benton River Press* 1881a:2). On June 16, 1881, only a month and half after the Opium Act took effect, the sheriff in Fort Benton raided his first opium den. In this case, similar to that discovered by archaeologists at Big Timber, the opium den was located within a Chinese laundry on Front Street. After conducting surveillance on the laundry for three days, the sheriff, with news reporter in tow, raided the establishment and arrested four Chinese men, including the owner of the building, Ah Gow. Three of the four Chinese men claimed that they did not actually smoke opium,

Figure 9: Opium can recovered from German Gulch, scale in mm. (Photo by author, 2009.)

though one of them was found with the pipe still in hand during the raid (*Benton Record* 1881a:2).

During the subsequent search of the building the sheriff discovered accoutrements that provide a glimpse into the materials found at a Montana opium den: a dozen ceramic opium pipe bowls, three brass stems, opium lamps, weighing scales, two "tins" of opium weighing a pound each (Figure 9), and a glass jar and several small glass bottles also full of opium (*Benton Record* 1881a:2). Using the prevailing costs of opium from Helena at $15–$17 per pound, the value of the seized opium probably amounted to well over $40 retail, a substantial drug bust in 1880s Montana. According to the report Ah Gow was charging about 50 cents per smoke of opium, twice as high as prices reported in Wylie and Fike (1993:262).

It also appears that the opium seized by the sheriff had bypassed the duty tax during importation, adding to the charges and costs assessed to the Chinese men (*Benton Record* 1881b:2). After several days in Fort Benton's jail the four Chinese men were arraigned before Judge Tattan. The three customers of the opium den were uniformly fined $5 and costs, but the proprietor was charged $50 or up to two months in jail. One of the Chinese men arrested refused to pay his fine, as he "has a sore hand

and can't do any washing anyway, so he thinks it a matter of economy to board with the county a few weeks. But he misses his opium and suffers terribly in the consequence" (*Benton Record* 1881c:2).

Fines associated with these cases ranged widely, based on costs of the court. Oolong, a Chinese man arrested for opium smoking in 1881, was fined $35 by the Butte City Court (*Helena Independent* 1881g:3), while an unnamed Chinese man and woman in Butte the same year received fines of $30 each (*Helena Independent* 1881h:3). December 1884 raids on several supposed opium dens in Virginia City netted four Chinese opium smokers, and they were convicted and assessed a fine of $22 each (*New Northwest* 1884:3). The operator of an opium den in Missoula arrested in April 1882 received a fine of $15, while the eight others served jail time in lieu of a fine (*Weekly Missoulian* 1882a:3). In Fort Benton three convicted Chinese opium smokers received a fine of only $5 each as punishment (*Benton Record* 1881a:2). The sale of confiscated opium smoking paraphernalia was also a normal part of prosecuting of these cases (*Helena Independent* 1881g:3, 1882f:3). It is odd that cities such as Butte and Helena would resell opium smoking paraphernalia, as only other opium smokers would presumably purchase such accoutrements unless as a novelty.

An arrested individual unable to pay the assessed fine and court costs was lodged in jail for a set amount of time (*Helena Independent* 1882b:3, 1882d:3, 1882f:3; *New Northwest* 1884:3). Stays in jail by the convicted party would vary by the assessed fine and community of the committed crime. Three Helena Chinese convicted of smoking opium in February 1882 were unable to pay their fines and spent 14 days in jail (*Helena Independent* 1882d:3). In another instance, the sale of confiscated paraphernalia of another convicted Helena Chinese opium smoker cut his stay to only five days in jail (*Helena Independent* 1882g:3). In a clear abuse of the Montana territorial legal system, five unnamed men in Livingston "were summoned as witnesses in the preliminary trial of a Chinaman for keeping an opium joint, swore that they knew all about it, their object being to be held in jail as witnesses till the next term of court, three months hence" (*Helena Independent* 1884a:3). As they were unable to pay bail and were held for free room and board for the three months, they accrued more than $900 in witness fees.

The assessed success of the Opium Act and its prosecution varied between locations. In the waning months of 1881 some editors believed that the Chinese were still unaware that opium smoking was illegal in

Helena but that in Butte they were intentionally defying the law (*Helena Independent* 1881f:3, 1881j:3). A reporter writing about the Helena Chinatown in 1881 noted that "there are no opium dens as of old, or if there are, the fact is kept secret" (*Helena Independent* 1881e:3). This article likely demonstrates a shift in the open-door policy toward the European American population by the Helena Chinese and their opium smoking establishments to avoid criminal prosecution. Moreover, this pattern is supported by accounts from Fort Benton (*Benton Record* 1881a:2). By the spring of 1882 reporters noted that opium-related arrests continued but on an occasional basis or less frequently than previous years. Correspondents noted that sufficient time had passed for the Chinese to make the practice more secretive and that "no law has yet been framed authorizing a forcible entrance into a dwelling, even though it be known that the place is frequented by personas addicted to the pernicious, soul destroying habit in question" (*New Northwest* 1882a:3).

As was pointed out by the *New Northwest* (1882a:3), there was little law enforcement could do to enter homes without significant evidence or actually witnessing business transactions within the public sphere. This legal issue spilled over into debates regarding Helena's City Charter in 1881. There was a cross-town difference of opinion between the *Helena Daily Herald* and the *Helena Independent*. Apparently the *Herald* staff supported a clause in the City Charter that would allow the raiding of suspected Chinese opium dens and the prosecution of Chinese individuals smoking in their own home (*Independent* 1881c:3), which is in fact the essence of the Opium Act. The *Helena Independent* staff felt that opium smoking was part of the Chinese culture, and opium was as "beer is of the German, wine of the Frenchman, and whisky of the American" culture, and its regulation was not the purview of a city (*Helena Independent* 1881b:3). Effectively the arguments centered on the private versus public consumption of opium. The *Herald* editors appeared to support a mandate to protect the Chinese from the perceived evils of opium, while the *Independent* editors focused on the sale of opium and related paraphernalia as the only true crime. While the argument focused on a single topic within the Helena City Charter, the debate itself illustrates the differing opinion on opium legislation and the effect of the drug on Montana society.

Montana's Opium Act provided a tool for prosecuting opium users in the territory. Similar to many local and national drug enforcement laws in historical times and the modern period, the biggest effect of the Opium

Table 11. Chinese archaeological sites and opium paraphernalia discovered

| Site # | Site Name | Pipe Bowls | Opium Cans | Fittings | Saddle |
|---|---|---|---|---|---|
| 24LC1083 | Marysville | X | X | X | — |
| 24LC1711 | Upper Cave Town | X | X | — | — |
| 24LC1723 | Upper Cave Gulch | X | X | — | — |
| 24MA0723 | Virginia City | X | X | — | X |
| 24MN0249 | Louiseville | X | X | — | — |
| 24MO0110 | Montreal or "Old Town" | — | X | — | — |
| 24PW1021 | Blackfoot City | X | X | — | — |
| 24SA0155 | Weeksville | — | X | — | — |
| 24SA0591 | West Noxon | X | X | — | — |
| 24SA0592 | Submerged Noxon | X | X | — | — |
| 24SA0593* | Thompson River | X | X | — | X |
| 24SA0594 | East Eddy | — | X | — | — |
| 24SA0595 | East Weeksville | — | X | — | — |
| 24SA0596* | "Last Chance" | X | X | — | X |
| 24SA0597* | "Fu Sang" | X | X | — | X |
| 24SA0598* | "Taklamakan" | X | X | — | X |
| 24SA0599 | Perma Bridge | X | X | — | — |
| 24SB0212 | German Gulch | X | X | X | X |
| 24SW0738* | Big Timber | X | X | X | X |

* = Artifacts also include some opium lamp parts such as the cage, glass, and wick supports.

Act was merely suppression of the public use of the drug, while clandestine use continued. Opium-related arrests continued in Fort Benton into the 1890s, and in 1900 a reporter described the large opium den within a restaurant on Front Street in Missoula (*Fort Benton River Press* 1890:2; *Daily Missoulian* 1900:3). Even rural communities like Whitefish in northwest Montana had arrests tied to the smoking of opium as late as 1923 (*Whitefish Pilot* 1923:2). In some cases the operation of opium dens was used as justification to force Chinese to leave the community, as was attempted in Livingston in 1887 (*Glendive Independent* 1887:2). As noted by Ahmad (2007:67) the failure of local and state anti-opium legislation spurred support for federal laws to combat the narcotic issue more adequately. Archaeological evidence demonstrates the ubiquitous use of opium in Chinese communities before and after the passage of the Opium Act.

Perhaps the most important point to take away from Montana's passing of the Opium Act and irregular enforcement was that for the first time in the territory's history, a law specifically targeting the Chinese passed, was enforced, and failed to be overturned by legal challenges. The failure of anti-Chinese legislation in Montana, such as the Alien Law in the early 1870s or the poorly enforced male-owned laundry tax had little overall impact on the Chinese community. While the law might not have fulfilled the desire of these supposed progressive European American Montanans to exterminate the opium habit, the Opium Act and the virulent debate surrounding it forever changed the tone of anti-Chinese legislation. Earlier territorial legislation like the Alien Law and laundry tax specifically targeted the economic causes of Chinese immigration, but now the discussion shifted to the perceived negative social impacts caused to Montana and surrounding states through the use and sale of opium. This shift in public discourse would lead directly to the overwhelming support of federal anti-Chinese legislation in the year following the Opium Act's passage.

## Northern Pacific Railroad

The arrival of the Northern Pacific Railroad (NPRR) to Montana Territory in the fall of 1882 and spring of 1883 was nearly 30 years in the making. Discussion of a northern railroad route connecting the Atlantic and Pacific oceans began in the early 1850s with the Mullan Road survey of the lands to become Montana Territory a decade later (Renz 1980:13). The first mention of the NPRR appeared in 1857, but the company was not established until 1864 when the federal government granted lands for its construction (Renz 1980:15). As signed into law in 1864 by President Abraham Lincoln, the United States federal government granted the Northern Pacific Railroad Company nearly 60 million acres of land, the largest such allotment in the nation's history for railroad construction (Smalley 1883:116; Renz 1980:19).

The proposed route of the NPRR would begin just west of Duluth, Minnesota, and stretch to Puget Sound in Washington, following the plan established by Governor Isaac Stevens and supported by evidence by Lt. John Mullan (Smalley 1883:77–84; Daniels 2000:55). Over the next 15 years there was sporadic activity on the physical construction of the NPRR due to mismanagement and a bad national economy in the 1870s. Even with the hurdles, the NPRR extended from Minneapolis to Bismarck by 1873 with the Pacific Division from Kalama to Tacoma completed that same year (Smalley 1883:196; Daniels 2000:55). In the mid-1870s the eastern end of the NPRR slowed and was only 18 miles west of Bismarck at the beginning of 1879, though crews quickly pushed that number to 100 by year's end (Renz 1980:65).

Under the direction of President Frederick Billings from 1879 to 1881, the NPRR planned the main routes to be used for the completion of the transcontinental line. More significantly, the Chinese side of the NPRR story began in earnest in 1879 with the final selection of a route from the confluence of the Snake and Columbia rivers northward to Lake Pend Oreille (Renz 1980:63). This section would become the Pend d'Oreille Division of the NPRR and had 300 Chinese workers involved in grading near Ainsworth, Washington, by late 1879 (Lewty 1987:52). Many of the initial Chinese laborers working for the NPRR and ancillary lines in Washington State were veteran railroad workers from the Central Pacific Railroad line completed ten years prior in 1869. While there were Chinese laborers on the NPRR's eastern half, most of their activity was from Missoula westward to the Pacific coast.

In early 1881 Henry Villard, head of the Oregon Railway and Navigation Company, successfully achieved a hostile takeover of the NPRR, which led to the resignation of former president Billings (Renz 1980:75–77). By late 1881 Henry Villard had become the new president of the NPRR (Renz 1980:78), and the next two years saw the finalization of the transcontinental line in an unprecedented expenditure of labor and capital. Villard weighed the needs of completing the line in a timely fashion with the current labor pool and subsequently began an aggressive campaign to import large numbers of Chinese workers from other areas of the United States and Canada, and even directly from China, successfully circumventing the Chinese Exclusion Act through powerful friends in the federal government. Over a decade earlier the Central Pacific Railroad had hired thousands of Chinese laborers, effectively cutting labor costs by a third (Wrobleski 1996:6). An editorial in the *New York Times* encapsulated the need for Chinese laborers, as "it is impossible to build railroads here without them, and the failure to secure as many as can be used . . . makes it difficult to push work as fast as railroad managers wish" (*New York Times* 1882:3). In a speech to Philadelphia investors in May 1882, Villard placed full credit for the recent prolific progression the NPRR firmly on the backs of Chinese labor (Renz 1980:89).

The NPRR generally paid the Chinese 50 cents less than their European

American counterparts, with the going rates at $1.00 and $1.50 per day, respectively (Lewty 1987:150). Swartout (1988:47) states that European American workers received twice the pay of the Chinese on the NPRR, though it is likely the numbers varied by location, tasks performed, and skills required. In a particularly unpleasant dry and dangerous section of the western half of the NPRR, managed by a subsidiary called the Oregon Railway and Navigation Company, Chinese laborers won a labor strike for an additional fifty cents, to raise their pay to $1.50 in 1882. The main reason the Chinese won their strike was that the area was so inhospitable that the contractor could not find enough willing European Americans to complete the work (Lewty 1987:117).

More significant for the Chinese, perhaps, was the growth of the credit-ticket system that sponsored their immigration. Throughout the mid-1870s American entrepreneurs traveled to China to secure cheap labor for the railroads, including the Northern Pacific. To facilitate this immigration, Chinese societies, both legal and illegal, sponsored the immigration of laborers but on the condition of repayment to the debt accrued for travel. Sadowski-Smith (2008:783–784) notes that this created the appearance of unfree labor, or even slavery, and helped to spur even more restrictive legislation on Chinese immigration, phrased for the benefit of protecting these workers. Regardless, at the wage rates the NPRR was paying the Chinese laborers, it would have taken a substantial time to work off the travel debt as part of the credit-ticket system. With upward of 135 Chinese loaded into freight cars and hauled to the front towns, the low status of the laborers within the NPRR fostered the growing public opinion that these immigrants were chattel property of the railroad and rich Chinese merchants (Lewty 1987:96).

As the western half of the NPRR moved toward the Montana border, the number of Chinese working on the grade continued to rise dramatically. Some estimates place 1,200 Chinese laborers on the Oregon Railway and Navigation Company grade along The Dalles in Washington in June 1881 (Lewty 1987:84). The transition to a heavy dominance of Chinese laborer continued through late fall of 1881 and into 1882. For example, of the 917 laborers on the grade between Spokane Falls, Washington, and Lake Pend d'Oreille, Idaho, 656 were Chinese, making up over 71.5% of the total workforce. According to the labor statistics, in this section of the grading there were 635 Chinese laborers under the direction of 21 Chinese foremen, for a rough estimate of about 30 graders per supervisor (Renz

1980:82). These numbers do not take into account any Chinese crews overseen by European American foremen. The numbers pale in comparison to the laborers required to undertake construction of arguably the most difficult section of the NPRR, stretching from Cabinet, Idaho, to Helena, Montana.

The section of the railroad from Lake Pend d'Oreille eastward to Helena was named the Clark's Fork Division (Renz 1980:98). Between 1879 and early 1882 only the stretch of grade from Sandpoint, Idaho, to Perma Bridge, Montana, was considered the Clark's Fork Division, but after determination was made to run the NPRR through Helena instead of Butte in 1882, the remainder of the line became that designation (Smalley 1883:415). J. L. Hallett was named the division boss in late 1881 and oversaw final engineering and survey work along the Clark Fork River from Idaho to Perma, Montana. During the winter of 1881–1882, the railroad camp at Cabinet Landing became the lead of the construction effort. Significant blasting and tunneling were required to pass eastward out of Cabinet Landing and into Montana Territory, and most of this work was done by Chinese laborers. Almost overnight Cabinet Landing became a boomtown with hundreds of dwellings, 35 saloons, two Chinese laundries, and a store supplying goods for the Chinese laborers but operated by a European American commissary (Renz 1980:83–84). Incredibly, during the late winter of 1881 and early spring of 1882, Cabinet Landing supported 2,600 Chinese laborers and 1,400 European Americans (Renz 1980:84). Lewty's (1987:93) estimates put the population at Cabinet Landing slightly lower, with 2,300 Chinese and 1,250 European Americans. Regardless, to put these numbers into context, the Chinese population of Cabinet Landing alone was several hundred people more than the entire Chinese population of Montana Territory at the same time. This large aggregation of railroad workers became euphemistically known as Hallett's Army by the Montana press.

By July 1882 Cabinet Landing crews abandoned the area, with newer settlements in Montana springing up at the advancing front of Hallett's Army. According to Renz (1980:84), "these front towns were even more transient than the mining camps." In 1881 crews completed a wagon road from Thompson River (just east of Thompson Falls, Montana) to Missoula (Renz 1980:84), and after the snow melt they finished the road from Thompson River to Cabinet Landing (Lewty 1987:93). This road served as the main route for supplying the front crews with provisions and construction supplies until the track caught up to the end of the line camps. In July 1882 there were an estimated 3,900 to 4,000 Chinese laborers working between Noxon and Trout Creek, Montana, nearly twice the over-winter population in Idaho (*New Northwest* 1882b:3; Lewty 1987:94–95). Some newspaper estimates placed the total number of NPRR workers in Montana in August 1881 at 10,000 (*Helena Independent* 1882h:3). A proportion of the new western front community moved from Cabinet Landing to Rock Island in Montana, though even by late summer this bustling settlement disappeared.

Research by avocational archaeologist Gary Weisz located the remains of nine Chinese railroad worker camps associated with the Northern Pacific construction. These camps stretch from Noxon to Perma Bridge, Montana, a distance of over 100 miles. Further information on the material culture of these camps is included in chapter 4 and in Merritt and colleagues (2012). The two camps near Noxon, Montana, were likely constructed in spring 1882 and only occupied a short time before abandonment. It is likely that Chinese and European American maintenance crews occupied these settlements into the 1890s, though this is hard to determine without further evidence.

Some of the most intensive work for the NPRR's western half in Montana occurred between Thompson Falls and Perma, Montana, with two river crossings and major blasting work west of Horse Plains. According to archaeological evidence there were substantial railroad settlements at the mouth of Thompson River, where it flows into the Clark Fork, and at least four additional camps east of Paradise (Weisz 2003). At the Thompson River camp, also called Ashleyville, occupied during 1882 by the NPRR there were distinctly separate areas of Chinese and European American encampments. Weisz (2003) noted the presence of at least a dozen tent platforms in the Chinese section, and archaeological materials indicate that this was indeed a major camp for both groups. Fullerton (1961) noted the presence of several exhumed Chinese graves at the Thompson River camp. By late summer of 1882 the Thompson River camp was abandoned with front work heading toward Horse Plains and the NPRR towns of Eddy and Weeksville.

"Bad Rock"—a large rock outcrop that extended into the north bank of the Clark Fork River—was between Thompson River and Weeksville, and the NPRR needed to remove this obstruction. This outcropping had obstructed most human travel along the Clark Fork for centuries (Helterline 1984:3–4). According to Helterline (1984:33), Chinese did most of the blasting for the Bad Rock project, and many died as a result of these efforts. Chinese laborers also worked on portions of the required blasting for the 1860s construction of the Central Pacific Railroad in the Sierra Nevada Mountains of California (Dirlick and Yeung 2001:92). Even today Bad Rock continues to be a problem for transportation; the Montana Department of Transportation had projects in 2005 and 2006 to blast additional sections of the outcropping without success.

As the NPRR front crews moved farther eastward, providing these crews with adequate supplies became more difficult. In October 1882 the companies contracted to provide room and board for the NPRR crews at the grading front, raised their weekly charge from $4 to $5 (Lewty 1987:98). In response European American NPRR employees attempted to drive all the Chinese from the front camps in hopes of removing the competition for supplies. In addition some of the European American workers took control of the Auditor's Office in Montana Territory along the NPRR grade (*Idaho Tri-Weekly Statesman* 1882:1). NPRR quickly regained control of the situation and discharged several hundred of the European American workers involved in the unrest (Lewty 1987:98).

Weeksville was located approximately seven miles east of Bad Rock and was a major front town for the NPRR. Weeksville received more historical attention in the newspaper accounts than most of the other temporary railroad camps during this time, including Eddy, just two miles away. Both these settlements boasted substantial Chinese populations as the railroad progressed through the Clark Fork Valley in the summer and fall of 1882. Weisz (2003:218) notes that archaeological evidence suggests most of the Chinese section of the settlement was in what is colloquially known as East Weeksville, while the European American workers and businesses were in the main part of town several hundred yards distant. It appears that Weeksville also boasted at least one Chinese laundry, as in 1883 it was robbed of money and clothing by a group of European American thugs (Helterline 1984:30)

Weeksville and its occupants had a rough and tumble reputation, with at least several dozen saloons. First initiated as a logging town to supply ties to the NPRR construction sites, Weeksville became one of the main over-wintering spots of Hallett's Army as the tracks did not reach the town until January 1883. Locals nicknamed the town "Bloody Weeksville" due to the number of murders in late fall of 1882, but the organization of vigilance committees in December that year quieted the roughest elements (Hagerman-Benton 2007:13). The vigilance committees, in pursuit of their mandates, lynched 10 men in February 1883 to express the point of law and order (*New York Times* 1883:3). NPRR crews largely abandoned Weeksville in February 1883 as the front moved to Horse Plains and areas east toward the Flathead Indian Reservation (Helterline 1984:32).

As Hallett's Army approached the Flathead Indian Reservation from the west, the Indian agent, Peter Ronan, sent an urgent telegram to the Bureau of Indian Affairs in Washington DC. While the NPRR had acquired permission to build the railroad through the Flathead Reservation, Agent Ronan feared that the less savory characters traveling with Hallett's Army would wreak havoc on the alcohol-free reservation. According to some reports there were 6,000 Chinese and 1,400 European American laborers employed on Hallett's section, with "1,000 loose women and men hanging on as the ragged edges of this army" (*New Northwest* 1882c:3). In short order Agent Ronan was able to negotiate a settlement that put to work crews of Mormons on the reservation.

Hallett's Army continued the grading and track laying from Plains to the border of the Flathead Indian Reservation throughout the winter of 1882–1883. There are few reports and little historical documentation for this section of the Clark's Fork Division of the NPRR, though archaeology can provide some illumination. Between Paradise and the third crossing at Perma, there were at least four NPRR camps occupied by the 6,000 Chinese and 1,400 European American workers. The only one of these camps with a historically known name was "Last Chance," which was located on the bank of the Clark Fork River, just downstream from its confluence with the Flathead River. This location was the last chance for Hallett's Army to acquire liquor and prostitutes before moving too close the Flathead Indian Reservation, hence the camp's moniker (Renz 1980:95).

Weisz (2003) located three NPRR camps east of Last Chance toward Perma, Montana, consisting of tent platforms and artifact scatters. Another camp two miles upstream, and smaller than Last Chance, might be associated with a blasting crew working in front of the main grading crews. Wrobleski (1996) discovered similar small satellite camps that were distant from larger aggregations along the Virginia City and Truckee Railroad in Nevada, probably representing small maintenance or specialty crews. The next camp appears to have been a main habitation for the Chinese graders and lacked significant evidence of any European American occupants.

Figure 10: West Portal, Mullan Tunnel. (Photo by author, 2008.)

The last NPRR construction camp recorded along the Clark's Fork Division is on a river terrace just before reaching the Perma Bridge. By April 1883 Hallett's Army completed the grading, track laying, and bridge at the end of the line at Perma. On April 7, 1883, the NPRR discharged Hallett's Army, which had shrunk to 900 Chinese and 800 European American laborers, and most workers traveled to Oregon to complete additional spur lines (Renz 1980:96). What happened to the 6,000 Chinese who were reportedly approaching the Flathead Indian Reservation in December 1882 is unclear, but it is likely that the NPRR reassigned the bulk of the crews to more pressing areas for construction, such as the Oregon spur lines, or even the NPRR grades east of Missoula and west of Bozeman. According to Renz (1980:98), by June 1883 Chinese crews were grading in the Hellgate Canyon east of Missoula, as the eastern half of the NPRR pulled into Helena.

Just west of Helena, NPRR crews worked tirelessly tunneling through Mullan Pass in the fall of 1882. Thirty years before, crews working for Lt. John Mullan had surveyed and graded a military road over this same pass, and the path became one of the major transportation routes between eastern and western Montana Territory. Taylor and Taylor (2002) do not discuss the ethnic composition of the workforce engaged in constructing the Mullan Tunnel during 1882–1883, and the data behind Swartout's (1988:47) assertion of the role of Chinese labor in its construction remain undiscovered. Swartout (2001:380) clarifies that his evidence for the Chinese labor force on the Mullan Tunnel project came from correspondence within the NPRR. A famous photo showing what appear to be Chinese workers at the western face of the Mullan Tunnel is generally attributed to around the period of completion of the tunnel in 1882–1883. According to historical documents the front façade that includes the "Mullan Tunnel" inscription was not added until the 1890s. That photograph likely dates from the 1890s through early 1910s and might indicate a reunion of workers, as suggested by Taylor and Taylor (2002:41), or it may simply represent an advantageous photo opportunity by Chinese tourists. In the summer and fall 2009 (Figure 10), Montana RailLink demolished both facades of the Mullan Tunnel in order to widen the existing cutting (Kidston 2009:1A).

While crews finished the boring of the Mullan Tunnel in the early fall of 1883, there was a race between the western crews, led largely by Chinese workers, and the eastern branch, comprising mostly Irish and other European American laborers, to the desired finish point near Gold Creek, Montana. Only 10 miles of the NPRR remained uncompleted in the morning of August 23, 1883. On a signal, crews from the west and east, each five miles from the finish flag, began the final grading and track laying.

At 1:50 p.m. the crew from the east reached the finish line with the western crew still nearly a mile away. Renz (1980) does not mention the ethnic makeup of the two construction crews during the August 23 race, though Renz's (1980:98) indication of the western crew's composition in June 1883 suggests that at least some Chinese were involved in the final race.

Between August 23 and September 8, 1883, NPRR construction crews completed a bypass track around the spot of the final ceremonies. Once the bypass was completed, the crews tore up one mile of track around the newly constructed pavilion. Henry Villard and hundreds of invited guests arrived at Gold Creek, Montana, to witness the driving of the final spike on September 8 (Renz 1980:267). According to a translation of a German journalist's reports on the day of the final spike ceremonies, a group of Chinese competed with European American laborers to finish the last mile (Lindau and Trautmann 1985:65). The original narrative by the German reporter illustrates the excitement of the last race:

*Who would finish first? Rails slammed onto ties, taking their places as if by magic, and were fastened with spikes hit in a violent rat-a-tat-tat. Suddenly a wild, feverish howl! Breathless and bathed in sweat, sunburned and dusty fellows reached the goal and screamed and yelled and waved their hats. Victory from the east—the Caucasians! The Chinese gained the spot five minutes later. (Lindau and Trautmann 1985:65)*

This is the only known mention of the ethnic composition of the construction crews at the final ceremonies and at least provides some credit to the Chinese laborers, though in the end the European Americans won the contest due to the delays caused by a derailed hand cart on the western half of the race (Renz 1980:267). Renz (1980:267) notes that J. L. Hallett was in charge of the western crews, and his heavy use of Chinese laborers in the NPRR construction efforts of western Montana support the possibility that at least those crews had some Chinese.

Nolan (1983:9) notes that the final spike celebration had photographers and newspaper reporters in great number; though it appears none captured a single representative of the nearly 10,000 Chinese employed by the NPRR for its construction. The ethnic cleansing of the photographic and historical record of the NPRR is similar to that in the completion of the Central Pacific Railroad in 1869, where the Chinese were excluded from the official photographs (Pegler-Gordon 2006:51). Pegler-Gordon (2006:51) asserts that the exclusion of the Chinese from the official documentary record of railroad construction is a microcosm of the treatment of that group's role in the history of the United States. A famous painting by Amedee Joullin, commissioned by the NPRR and prominently displayed at the Montana State Capitol, shows President Grant, Henry Villard, Crow Indians, and other dignitaries but pointedly omits the Irish and Chinese laborers who supplied the bulk of labor to the project (Lambert et al. 2002:50–52)

After the festivities on September 8, 1883, the majority of the NPRR workmen were released from the main line, and many of the laborers went to Washington and Oregon to assist with those sections. A year after the festivities the death toll of the NPRR's construction was tallied, and estimates placed the number of buried Chinese along the line at 1,000 (*Helena Independent* 1884b:3). Lewty (1987:96) notes that as the backbone of the NPRR construction effort, the Chinese suffered most from the work as they were killed in falls, explosions, and some even from scurvy due to poor diet afforded them on the front. Some of the most dangerous work was the blasting of rock faces between Cabinet Landing, Idaho, and Missoula, Montana (Hagerman-Benton 2007:13).

Helterline (1984:33) discusses several blasting incidents, including one that "hurled an entire point of rock into the coolies' campsite, killing and maiming upwards of 150 at one annihilating blow." In June 1883 approximately 24 Chinese died and nearly 50 were injured when a construction train collided with a lumber train near Heron, Montana (*Spokane Falls Review* 1883:1). According to the newspaper reports, over 80 Chinese laborers were sitting atop open flat-bed cars heading westbound toward Idaho when an unscheduled lumber train came around the corner, leading to a head-on collision. The graphic description of the incident highlights the harsh realities of NPRR construction: "Men were torn to pieces, limbs were crushed to pulp, and bodies deformed beyond recognition" (*Spokane Falls Review* 1883:1).

The only officially recorded Chinese graves along the NPRR grade in Montana are within the boundaries of the Thompson River camp. In 1961 local amateur historian Neil Fullerton interviewed several Noxon and Heron locals who provided some anecdotal stories about the Chinese, including the location of three grave sites east of Thompson Falls. Fullerton (1961) noted that rocks encircled all three graves, and their arrangement suggested disinterment. Two of the graves are oriented due north-south, while the third grave, twenty-five feet away, is oriented east-west. Attempts to relocate the graves in 2008 were inconclusive, though the dense undergrowth and heavy mosquito presence prevented detailed surface survey.

There are likely numerous other grave sites along the old NPRR grade, as evidenced by other local histories. According to Helterline (1984:33), there were numerous one-foot deep graves along the NPRR right-of-way. At Heron most of the Chinese who died during construction were buried in an orchard, while at Noxon they were interred near the Clark Fork River. In the 1920s a local resident discovered several Chinese graves on a gravel bar on the north side of the NPRR line near Noxon. Just before completion of the Cabinet Gorge Reservoir in the 1950s, many Noxon residents exhumed Chinese bodies from a cottonwood-shaded plot near the NPRR depot and reinterred them in the city cemetery on Pilgrim Creek (Leeson-Vanek 1996:117).

Besides the dangerous work of the construction, the construction camps were dangerous places as well. As already mentioned, Weeksville quickly gained a reputation as a rough and rowdy town and saw the robbery of a Chinese laundry in 1883. It appears that the Chinese were not docile victims of the harsh treatment by their European American employers. In July 1882 a European American section foreman named Meagher, who was in charge of a Chinese work detail, was found murdered in his tent. Without formal due process a gang of Meagher's friends enacted their vengeance upon the Chinese, killing six and wounding 13 without ascertaining their involvement (*Helena Independent* 1882f:3). A similar incident occurred a few months later in January 1883 when Larry Herne, another European American section foreman in charge of Chinese crews, was found dead in his tent at Weeksville. According to the newspaper accounts, it appears that the Chinese under his charge killed him with a hatchet while he slept (*Helena Independent* 1882j:3). While both these stories hinge on the accurate investigation and reporting of the crimes, it seems plausible that the Chinese might have conducted these actions.

The last major recorded act of violence along the NPRR was committed within the Chinese community at Horse Plains in 1883. According to the newspaper reports two Chinese men killed a Chinese paymaster and robbed him of $3,000 at Horse Plains and, during their flight, injured another Chinese man who belonged to the Wing Sing Company (*Helena Independent* 1883a:3). Over the summer of 1883 three Chinese men captured the two assailants, and after trial they were hanged in Missoula on August 16, 1883 (*Helena Independent* 1883d:3).

There were a few Chinese settlements along the NPRR through the remainder of the 1800s and first two decades of the 1900s. Employed as track maintenance crews, the Chinese lived in wooden section houses at Noxon, Belknap, and White Pine. Belknap, a small settlement established along the NPRR line in the 1880s, was completely destroyed by the great 1910 fire that scorched millions of acres of Montana and Idaho. During the mid-1880s, however, Belknap had a Chinese laundry (Draszt 1998:3). Fullerton (1961) collected an oral history that recounted the presence of Chinese section hands at the White Pine NPRR Depot through the 1910s, and other local accounts support this assertion (Draszt 1998:71). Recorded as 24SA0568, the White Pine mercantile and the remnant foundations of the depot and Chinese section hand shack are owned by private parties and might retain significant integrity.

In the mainstream history books on the NPRR's construction (Smalley 1883; Renz 1980; Lewty 1987; and Taylor and Taylor 1999, 2002), it is clear that the Chinese contribution has been largely ignored by scholars and the public. As noted by Pegler-Gordon (2006), there was an intentional historical omission of the Chinese contributions from the literature in the 1800s, and this attitude has not dramatically changed since. Over the four years of the NPRR's peak construction phase, 1879–1883, the Chinese contributed well over two-thirds of the labor on the entire line, potentially hitting a population of 10,000. It is clear that the NPRR was critical to the rapid economic development of Montana through the 1880s, and thousands of settlers used the rails over the next four decades to settle the state's expanses. After completion of the NPRR, many of the Chinese workers left for the Canadian Pacific Railroad, completed in 1885. In the United States much of the NPRR maintenance fell to Japanese laborers who were not excluded by the Chinese Exclusion Act. For instance, Livingston, which served as the NPRR's major hub in south-central Montana, had three times as many Japanese railroad workers as Chinese counterparts in 1900. The only remnants of the Chinese contribution left today are less than a dozen archaeological sites along the old NPRR grade west of Missoula to Idaho and, of course the grade, rails, and blasted rock faces that are silent reminders of this forgotten history of thousands.

The next two trans-Montana railroads, the Great Northern and Milwaukee, used a majority of Japanese workers in their construction due to the effects of the 1882 Chinese Exclusion Act. Railroad companies required cheap labor for construction and maintenance, and with the inability to acquire Chinese workers, many managers imported Japanese laborers by the thousands. Between 1891 and 1907 labor contractors working for a variety of American businesses, including railroads, helped to import over 60,000 Japanese laborers (Ichioka 1980:326–327). Completed in the 1890s, the Great Northern Railroad went across Montana's northern boundary and employed several thousand Japanese laborers, some with Chinese foremen, in addition to a significant number of Irish and Eastern Europeans. The Milwaukee Railroad, Montana's first electrified railroad, used significantly fewer Japanese laborers than the Great Northern and was completed in the early 1900s (Ichioka 1980:325–326).

## Exclusion and Montana's Boycott(s)

At the peak construction of the NPRR, the United States federal government passed the Chinese Exclusion Act in 1882 in response to mounting pressure from labor unions and western legislative bodies. This was the first law passed in the United States that directly excluded a particular nationality from immigration, and it spurred additional anti-Chinese legislation in 1892 and 1902 (Gyory 1998:254). In its original 1882 form, the Chinese Exclusion Act banned any immigration of Chinese laborers into the United States for a period of ten years and exempted any individuals already residing in the country before its passage (Gyory 1998:261). This law effectively closed the doors to further Chinese immigration, which in turn precluded any potential for a robust and growing immigrant population in Montana. Yet in Montana during this period the Exclusion Act was covered in the local newspapers with little fanfare or vigor. The reason for this seemingly apathetic attitude remains unclear, but even so, public sentiment had now fully turned against the Chinese of Montana. As in other western states after the passage of the Exclusion Act, Montana's Chinese experienced an increase in violence and economic restrictions.

After passage of the Chinese Exclusion Act, there was an increase in anti-Chinese violence and boycotts across the western United States. During 1885 mobs swept through Rock Springs, Wyoming's Chinese quarter, resulting in an estimated 28 dead, 15 injured, 75 homes burned, and a forced exodus (Crane and Alfred 1940:49; Daniels 1979). By far some of the most brutal and horrific anti-Chinese violence occurred in Los Angeles, Denver, Seattle, and Tacoma between 1871 and 1886 (Schwantes 1982:373–376; Zhu 1997:146; see also Jew 2002 and Rusco 2002). Similar to Idaho (Zhu 1997:146), Montana did not witness such widespread or organized violence against the Chinese as in other parts of the American West. Zhu (1997:146) points out that in 11 anti-Chinese riots in Idaho from 1863 to 1900, 25 Chinese were killed and an additional 528 displaced. It seems that in Montana the violence was limited to singular incidents of malice, with most efforts focused on boycotts and legal restrictions sponsored largely by labor unions and especially the Knights of Labor.

At the start of mid-1880s anti-Chinese agitation in Montana, unknown parties demolished a Chinese laundry in Anaconda, killing three (*New Northwest* 1885b:3; *New Northwest* 1885c:3). Since Anaconda was headquarters for the Knights of Labor's efforts to provoke statewide anti-Chinese boycotts, it is supposed that there was some association with this movement. According to newspaper reports, unknown parties placed two large powder charges underneath and adjacent to a Chinese laundry on Main Street. The charges exploded at 3:00 a.m., destroying the laundry and shattering windows of businesses within 200 feet of the structure (*New North-*

*west* 1885c:3). The act was a clear signal to the Chinese, but who sent the message remains unclear, as no individual was ever charged in the crime.

By the late summer of 1885 the anti-Chinese agitation in Montana reached its peak, feeding off the events in Rock Springs, Wyoming, and Tacoma, Washington. In Butte the *Miner* and *Inter-Mountain* newspaper offices barred their windows and prepared for a potential violent riot similar to those in other areas of the West (*New Northwest* 1885e:3). A Knights of Labor–sponsored demonstration and rally in downtown Butte occurred in September 1885, with some estimates placing the crowd at 4,000 to 5,000 strong. At the rally many prominent local politicians spoke about the evils of Chinese labor depriving the working American of wages and employment, and speakers pushed for a complete boycott of all Chinese-operated establishments and any business that employed Chinese people (*New Northwest* 1885f:3). In quick succession other Knights of Labor assemblies supported the boycott, and by the end of 1885 Dillon, Walkerville, and Garrison joined the effort (*New Northwest* 1885g:3, 1885h:3, 1885i:3).

The Knights of Labor set the tone of the boycott, and in the September 1885 rally the leadership stated that they did not support the use of violence against the Chinese (*New Northwest* 1885f:3, 1885h:3), as they likely saw the drawbacks of such activity for garnering widespread and permanent support (Flaherty 1987:47). The Dillon Assembly of the Knights of Labor expressed their sentiment in a printed resolution that also addressed the Chinese:

> *To Chinese: Wishing you no harm personally, we advise you to listen to the voice of the people as heard in Oregon, Washington, California, Wyoming, Montana and other parts of the West, and consult your best interest in this matter. The Chinese must go. (New Northwest 1885g:3)*

It appears that in most instances, except perhaps the Anaconda laundry explosion and potentially unreported instances of violence, the Knights of Labor were successful in sponsoring a mostly nonviolent boycott of the Chinese. Other areas took the boycott further, as in 1885 Neihart, a hard-rock mining town in the Little Belt Mountains, passed an ordinance forcing the removal of Chinese from that community (*River Press* 1885:2). Another hard-rock camp called Maiden followed suit in 1886 with an anti-Chinese ordinance, and European American locals forcibly removed those Chinese who refused to leave (Robison 2009a). In 1887

Great Falls passed a similar ordinance that was still in effect in 1903 and involved the violent expulsion of some Chinese individuals, including dumping them into the Missouri River in a burlap sack (*Great Falls Tribune Daily* 1903:10). Residents of Livingston also proclaimed that the Chinese must leave that community in 1887, though there was no physical violence reported (*Glendive Independent* 1887:3).

Not all the local newspapers supported the boycott, as again the *New Northwest* stood up for the Chinese in their own peculiar way, noting that the presence of this population was an annoyance but a "necessary evil" due to the lack of a large pool of cheap labor (*New Northwest* 1885f:3, 1885h:3). Editors of the paper continued that if the Chinese left under duress there would be no one to replace them in laundries, gardens, or domestic service (*New Northwest* 1885d:3). It appears that the *New Northwest* supported the Chinese only under the auspices of existing treaty rights and because of the dangers of the spread boycotts against non-Chinese (*New Northwest* 1885h:3). In the *New Northwest*'s (1885h:3) opinion, this was "a dangerous encroachment on the personal and individual rights of the American citizen" and was "unrepublican, undemocratic, un-American."

In fact the *New Northwest*'s worry about the potential spillover effects of the boycott reached their own door in late October 1885. Due to the editorial support of the Chinese in the *New Northwest*, the Garrison Assembly of the Knights of Labor sent flyers to other assemblies urging them to boycott the newspaper. Instead of retracting their statements, the editors ran a story urging members of the Knights of Labor to pay up their outstanding debt and cancel their subscriptions, as "we have nailed our flag to the mast and ask no favors" (*New Northwest* 1885i:3). It appears that in the short term, cooler heads prevailed, the anti-Chinese boycotts of the mid-1880s suffered from widespread apathy from Montanans, and the sporadic activity failed to generate large-scale change in the demographic composition of the state.

Even with concerted efforts to remove them, the official Chinese population in Montana peaked in 1890 (Table 12), with 2,532 recorded across 16 counties (Federal Census Bureau 1896:42). The Chinese Exclusion Act and repressive labor practices in various industries led to a slowing of population growth for this cultural group by cutting off the new flow of male migrants and continuing the prohibition against women. This total put Montana in position as having the sixth largest Chinese population in the United States behind California, Oregon, Washington, New York, and Nevada (Table 13). Due to the loss of the 1890 census in a warehouse

Table 12. Montana Chinese population by county, 1890 Federal Census

| County | Chinese Population |
|---|---|
| Beaverhead | 92 |
| Cascade | 23 |
| Choteau | 42 |
| Custer | 18 |
| Dawson | 4 |
| Deer Lodge | 438 |
| Fergus | 9 |
| Gallatin | 39 |
| Jefferson | 46 |
| Lewis and Clark | 602 |
| Madison | 155 |
| Meagher | 37 |
| Missoula | 405 |
| Park | 23 |
| Silver Bow | 584 |
| Yellowstone | 15 |
| Totals | 2,532 |

Table 13. Chinese-born population by state, top six population centers. 1890 Federal Census, Bulletin 199, Census Bureau

| State | # of Chinese | % of Total Chinese in U.S |
|---|---|---|
| California | 72,472 | 67.4 |
| Oregon | 9,540 | 8.9 |
| Washington | 3,260 | 3.0 |
| New York | 2,935 | 2.7 |
| Nevada | 2,833 | 2.6 |
| Montana | 2,532 | 2.4 |

fire in the early 1900s, there is little historical evidence to determine the demographics of Chinese in Montana during this period.

In 1890 there were 107,475 Chinese in the United States, but by the end of the decade the population dropped by nearly 30,000 due to the strict enforcement of anti-immigration legislation. The proportion of Chinese to the total American population peaked in 1880 at 1.56%. By 1890 Chinese individuals made up only 1.15% of the United States population (Federal Census Bureau 1892:52). In the United States the 1890 census recorded a substantial decrease in Chinese population growth, which had increased only 1.91% from 1880. The previous two population increases were 80.91% from 1860 to 1870 and 66.88% from 1870 to 1880. The significant drop of Chinese population in the United States from 1890 to 1900 was a direct result of the renewal of the federal Chinese Exclusion Act, with the new version named the Geary Act.

When the Geary Act replaced the Chinese Exclusion Act in 1892, the new restrictions placed on the Chinese in Montana received substantial coverage in the local presses. The Geary Act not only renewed the provisions of the Chinese Exclusion Act but created new requirements for Chinese already living in the United States. Specifically, within one year after passage of the Geary Act, all Chinese had to register with the collector of local revenue for the district and receive a certificate of identity showing their name, age, local address, occupation, and photograph (Pfaelzer 2007:292). If a Chinese person refused, or could not register within that period, the individual was arrested and deported (Pfaelzer 2007: 300–301).

The Chinese Six Companies, based out of San Francisco, forbade any Chinese to register for the Geary Act in 1893 as an effort to create a united front against what they saw as an unconstitutional law. In certain areas the Six Companies enforced their will over other Chinese by promoting boycotts of merchants who registered, or by filing health complaints against restaurants and butcher shops (Pfaelzer 2007:300). By the original Geary Act deadline, only a fraction of the Chinese population registered with local agencies. After the failure of the original timeline, an amendment extended the deadline for registration. Many Chinese attempted to register due to this amendment and the defeat of a Supreme Court challenge to the constitutionality of the Geary Act. Non-Chinese populations in hundreds of communities around the United States now attempted to stop their registration by force, leading to dozens of violent and deadly episodes in 1893–1894 (Pfaelzer 2007: 307–315). Some estimates state that less than 10% of California's Chinese population ever complied with the registration for the Geary Act (Pfaelzer 2007:331).

In Montana, the initial period netted only a few Chinese registrants, though the extension of the Geary Act's deadline to May 1894 was more successful. In January 1894 the Chinese Six Companies, now accepting that they could not defeat the Geary Act, sent notices to all their branches and local leaders urging them to register in force (*Helena Independent* 1894b:3). By March 1894, Montana newspapers circulated that 2,000 Chinese already registered, with approximately another 2,000 yet to complete the process in the district (*Glendive Independent* 1894:2; *Helena Independent* 1894c:3). Montana, Idaho, and Utah comprised a single district, with many Chinese forced to travel several hundred miles to the closest federal registration station (*Helena Independent* 1894d:3). Some reporters felt that the elder Chinese in Montana refused to register, as deportation provided a free means of returning to China; they noted that the oldest registrant was 72 and the average age was 38 (*Helena Independent* 1894d:3). It appears that most Chinese in Montana and Utah complied with the Geary Act, with most of the district's shortfall coming from Idaho, where reports indicate that 1,000 failed to register (*Helena Independent* 1894d:3; 1894e:3). Fort Benton reported thoroughly on the number of Chinese registrants and noted that a local photographer did at least 21 of the portraits for the applicants (*Fort Benton River Press* 1894a:1; 1894b:1). At least some of the Chinese residents of Fort Benton used the studio of an African-American photographer, J. P. Ball in Helena.

During deliberations over the renewal of the Chinese Exclusion Act in 1891, Missoula's citizens initiated an anti-Chinese boycott (Morris 1997:117). Using Missoula's efforts as a template, by 1892 others were starting boycotts against Chinese establishments and workers in Montana after passage of the Geary Act, centered mostly in Butte and Anaconda and sponsored by labor unions (Morris 1997:118–120). The new boycotts of the 1890s brought interest from outside the state, as the *Salt Lake Herald* (1893:1) carried a supportive editorial on the efforts of labor unions in Montana. Attempted boycotts of 1891–1893 again failed miserably, due to the lack of consistent support (Flaherty 1987:36). Boycott of the Chinese also affected other non-union businesses that benefited from their labor, providing fertile ground for dissension. In contrast to the relative failures of the mid-1880s anti-Chinese agitation, the late 1890s brought a new vigor to these efforts and fomented the most successful boycott in Montana history, the Butte Union Boycott of 1896–1897.

There is no need to discuss in detail the 1896–1897 Butte Union Boycott against the Chinese, as other pertinent sources (Wunder 1980; Flaherty 1987; Schneider 2004) provide sufficient discussion. Not to be thwarted again, boycott organizers in Butte began a systematic and organized attempt in 1896 to drum up financial and moral support for the proposed efforts. By early 1897 the Silver Bow Trades and Labor Assembly officially supported the new boycotting efforts, bringing substantial finances and power to bear on the Chinese, and now Japanese, laborers in Butte (Flaherty 1987:38). The new Butte boycott effectively strong-armed local businesses that employed Chinese labor, as even a small break in business could destroy a small operation like a boarding house or restaurant

(Flaherty 1987:40–41). Chinese merchants like Dr. Huie Pock and Hum Fay immediately felt the effect of the boycott, with their business sharply declining in the first months of 1897 (Flaherty 1987:41). Some estimates suggest that at least 350 Chinese left Butte during the boycott due to their inability to locate work or sell their wares (Flaherty 1987:41).

Without the immediate support of the Chinese Six Companies in San Francisco, Dr. Huie Pock, Hum Fay, Dear Yick, and Hum Tong filed a lawsuit against 20 Butte businessmen active in the boycott (Flaherty 1987:41). The lawsuit specifically outlined that the defendants unlawfully deprived the Chinese of their ability to operate businesses and created a double standard in the legal protection of their ventures. Chinese plaintiffs hired William Fisk Sanders as their attorney, while the defendants employed Josiah and Melvin Wines. Interestingly, the judge for the Butte boycott case was Judge Hiram Knowles, who was no stranger to dealing with Chinese-related cases and who himself employed a Chinese domestic servant in his home. Judge Knowles placed an immediate injunction to stop the Butte boycott until after the completion of the trial, which did not begin until 1898 (Flaherty 1987:45).

In 1900 the case ended and Judge Knowles issued his determination that the boycott was unlawful and ordered the defendants to pay the Chinese legal fees accrued over the course of the trial (*Deseret Evening News* 1900:4; Flaherty 1987:46). The Chinese did not sue to recover any damages from the boycott, as they felt that such an undertaking would fail and would not end the efforts of the labor unions. After receiving the judgment from Knowles, the Chinese plaintiffs appealed to the United States federal government for payment of damages accrued between 1896 and 1900. Even with the support of the Chinese consulate in Washington DC, the secretary of state refused their plea, and no damages were ever awarded (Flaherty 1987:46).

The widespread boycotting efforts throughout the 1880s and 1890s even influenced the selection of a state capital. In 1892, due to the presence of Chinese labor, the Knights of Labor refused to back the proposal of Helena to become state capital unless the City Council supported the immigration restriction. The Helena City Council thus passed resolutions supporting the renewal of the Chinese Exclusion Act as proposed in the Geary Act (*Record-Union* 1892:1). Newby (1987) discusses how supporters of Helena's rival for state capital, Anaconda, seized upon the continued tolerance of Chinese in the city as a major black mark for voting. The anti-

Chinese efforts of labor unions found little support in Helena, due to the lack of substantial industries and related workers. Even African Americans used the Chinese population of Helena as one reason for supporting Anaconda as state capital (*Colored Citizen* 1894:1). It is worthy of note that the *Colored Citizen* was edited by J. P. Ball Jr., son of the African American photographer who took portraits of Chinese during the Geary Act period. Helena nevertheless won the bid for state capital in 1894.

Illegal immigration and human smuggling remained one of the largest problems associated with the continued exclusion of Chinese. As early as 1892 Montana Governor Toole requested federal monies to bolster the weak efforts to police the border with Canada (*Salt Lake Herald* 1892:1, 1893:1). By the early 1900s there were constant sweeps of northern Montana communities to discover the location of illegal Chinese immigrants. In 1904 immigration inspector Alfred Hampton located six illegal Chinese spread through Kalispell, Columbia Falls, Cutbank [sic], and Assiniboine (*Deseret Evening News* 1904:1). In 1930 a Whitefish man remembered smuggling alcohol and Chinese immigrants across the Montana-Canada border during the 1910s and 1920s (*Whitefish Pilot* 1930:3).

There were a number of Chinese deportations after 1900 in small communities like Fort Benton (*Fort Benton River Press* 1901a:1, 1901b:1, 1902:1, 1904:1), but most centered in the larger towns of Butte and Helena (*Helena Independent* 1933a:3). An attempted deportation of a Chinese man from Missoula facilitated a significant change in the way the United States applied deportation rules. A judge ordered the deportation of Foo Duck, the 23-year-old son of a Chinese merchant in Missoula (*United States v. Foo Duck* 1909). Foo Duck joined his father in Missoula in 1901, and during that time he studied English and worked as a cook and servant. A United States commissioner in Missoula deemed that Foo Duck was eligible to be deported due to his failure to have documentation under the Geary Act, and that he worked in the laboring class, which was the focus of the Chinese Exclusion Act. In appeal to the district court the judge found that Foo Duck could remain in the United States as he had immigrated to join his father and only worked as a cook and waiter after reaching adulthood (*United States v. Foo Duck* 1909:490–491).

During 1902 and 1914 legislators renewed the Chinese Exclusion Act in its various incarnations, though in 1924 the prohibition against immigration extended to 99 years, requiring no further renewals (Pfaelzer 2007:335). Chinese populations diminished in many Montana communities as a result of the boycotts, with all Chinese residents leaving some towns, like Anaconda (Morris 1997:123). Immigration law and the exclusion of new Chinese from coming into the United States and Montana forever changed the direction of this cultural group's future. It was clear that the heyday of Montana's Chinese community had passed, and that the future held little promise for their economic and social prosperity. The changing economic and legal landscape of Montana starting in the 1870s affected and coincided with the shift of Chinese labor from mining to a diverse array of service-based occupations.

# 3. Diversification, Collapse, and Aging, 1900–1943

## Chinese History and Archaeology 1900–1943

A rapidly declining and aging population and an aggregation of populations away from smaller settlements and into larger urban centers defined the early 20th century of Chinese experience in Montana. The collapse of placer mining in the late 1800s promoted the diversification of Chinese businesses in the first decades of the 20th century, as described in this chapter. The direct impact of the Chinese Exclusion Act and subsequent laws was the cessation of immigration, including the women and children needed to create a stable and growing population. The 1880s and 1890s signaled a significant change in the ways Chinese experienced the Montana frontier.

Before the 1880 Opium Act, the legal efforts to limit the freedom of Chinese immigrants in Montana proved generally unsuccessful. Changes to immigration law through the Exclusion Act successfully stemmed the flow of new Chinese workers to Montana and effectively halted any potential for significant growth in families. Illegal boycotts in Butte and other areas of Montana deeply affected the ways in which the Chinese conducted their businesses.

Simultaneous with the decline of Chinese population statewide beginning in the 1890s, the historical record of this cultural group increased. Better record keeping and the efforts of state and national forces to register and enumerate the Chinese population created a better overall image in the archival record during this later period. Unfortunately, the Chinese were entering the twilight of their time in Montana. While the Chinese did not entirely disappear by 1943, their population dwindled to a few hundred. Similar to patterns in Idaho (Fee 1993), Australia (McGowan 2005), and New Zealand (Ritchie 1986), the decline of Montana's Chinese population reduced the economic and social threat imagined by the host community. In the 1900s Montanans often viewed the Chinese as humorous, exotic, and in nearly all cases benign. While the warming of relationships between the

Table 14. Chinese archaeological sites in Montana relating to the 1900–1943 period

**Confirmed Chinese Archaeological Sites**

| Site # | Site Name | Period | Owner* |
|---|---|---|---|
| 24BE2186** | Blue-Eyed Annie | 1895–1901 | USFS |
| 24BW0497 | Lombard | 1890–1910 | BLM |
| 24GN0009 | Chinese Graves | 1900–1920 | USFS |
| 24GN0377 | Sierra Mine Site | 1870s–1910s | BLM/Private |
| 24GN0540 | Garnet | 1880s–1930s | BLM/Private |
| 24LC1032 | Forestvale Cemetery | 1898–1954 | Private |
| 24LC1083 | Marysville | 1870s–1900s | Private |
| 24MO0172 | Coloma | 1880s–1910s | BLM/Private |
| 24MO0266 | Fort Missoula | 1877–PRESENT | County/USFS/Private |
| 24PA1330*** | Huckleberry Camp | 1897–1900 | USFS |
| 24SA0599*** | Perma Bridge Railroad Camp | 1881–1883 | Private(MT Rail Link) |
| 24SB0212 | German Gulch | 1865–1900 | USFS/Private |
| 24SW0738 | Big Timber | 1882–1940 | Private |
| 24SB0765 | Butte | 1864–1940s | Private |
| 24SB0067 | Highland City | 1865–1870s | USFS |

**Disproved Chinese Archaeological Sites**

| Site # | Site Name | Period | Owner* |
|---|---|---|---|
| 24SA0122 | Poacher Gulch Terraces | 1900–1920 | USFS |

**Potential or Unconfirmed Chinese Archaeological Sites**

| Site # | Site Name | Period | Owner* |
|---|---|---|---|
| 24FH0446 | China Basin Cabin | 1920s–1930s | USFS |
| 24SA0568 | White Pine | 1881–1910 | Private |
| 24SB0742 | China Gulch Pads | 1870s–1920s | USFS |

*BLM = Bureau of Land Management, USFS = United States Forest Service

** = New site in 2008, no Chinese artifacts.

*** = New site discovered in summer of 2008.

Chinese and other Montanans was a welcome change, the interactions still focused on the immigrants as a quaint reminder of earlier times in the state.

Archaeological studies of sites reflecting this third period of Chinese heritage provide detailed information on their integration into 20th-century Montana. In particular, the most significant projects on 20th-century Chinese sites occurred in the last few years with Mitzi Rossillon's (2008) work in downtown Butte, and Justin Moschelle's (2009) excavation at Big Timber. There are currently 15 confirmed archaeological sites associated with this period and an additional three possibilities (Table 14). The Poacher Gulch Terraces, excavated in 2006 and 2007, also date to this third period and reflect more of modern Montana's mythic images of the Chinese experience than any actual historical or archaeological fact.

This final period ends with the repeal of the Chinese Exclusion Act in 1943 for strategic purposes by the United States government. The American government instituted a quota system for Chinese settlers in 1943, the tide of immigration had turned, and many social historians agree that this repeal signaled the start of a new era in Chinese-American history (Li 2006). Yet even this action failed to revitalize the Chinese population of Montana, and the numbers declined to their lowest level ever in 1950. The Chinese temple in Butte was torn down in 1943, and the last Chinese laundry closed in the early 1950s (Ciabattari 1976:16).

## Diversification of the Workforce

Chinese mining had steadily declined during throughout the 1880s and 1890s, due to changes in the availability of placer gold, restrictive legislation, and other societal factors. Most Chinese mining operations could not compete with the heavily capitalized ventures of their European American counterparts, and the restrictive federal Hard Rock Mining Law forbade them from undertaking any lode ventures. Thus the few remaining Chinese placer miners were a rare sight in Montana during the last decade of the 19th century, and their operations were usually small-scale and intermittent, worked by only a few individuals. University of Montana archaeologists discovered one such operation in 2008 on the Gallatin National Forest (Merritt 2010). In 1897 Wong Chong, president of the Quong Chong Mining Company, purchased a claim along Emigrant Creek approximately 11 miles upstream from Chico Hot Springs. This camp was significantly isolated from other Chinese communities; the closest substantial population living in Livingston, 40 miles away. At the site archaeologists discovered a few Chinese artifacts, but non-Chinese materials dominated the assemblage. This does not suggest that the Chinese of Emigrant Creek were more assimilated; instead it is likely due to the fact that they lacked access to their traditional goods because of their isolation.

By the 1900 census the Chinese population of Montana felt the effects of restrictive immigration legislation, and the population declined from a high of 2,532 in 1890 to 1,721. As in the preceding three decades, the Chinese population of Montana still clustered in the large mining districts of Silver Bow, Deer Lodge, and Lewis and Clark counties (Figure 11). Employment as miners accounted for only 7.3% of the workforce,

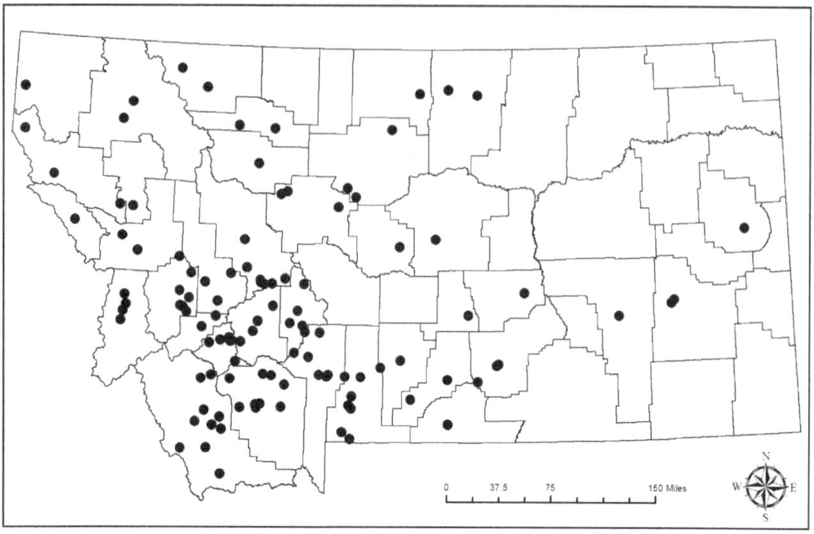


Figure 11: Chinese population centers in 1900. (Created by author.)

down from the 59.6% in 1880. Other industries superseded the mining sector, as the number of Chinese employed in the laundry, gardening, and domestic servant/cook areas rose sharply in 1900 (Table 15). The number of Chinese women listed as prostitutes also dropped considerably by 1900, with more now recorded as wives or homemakers (Table 16). In addition the legislation restricted the number of Chinese women in Montana, which declined from 78 in 1880 to only 36 by 1900, representing 4% and 2% of the total Chinese population, respectively. The male to female ratio increased throughout the 19th and early 20th century, with 14.23 men to one woman in 1870, 20.51 men to one woman in 1880, and an incredible 40.3 men to one woman by 1900.

As evidenced by the expansion of Chinese occupations recorded from the 1870 to 1900 censuses, there was a diversification of employment in the latter years of the 19th century, as seen elsewhere in the West (Lee 2006:5). While there were Chinese laundries, restaurants, cooks, and gardeners in the earliest days of territorial Montana, the proportion was relatively small compared to those involved in placer mining. As Montana's mineral economy quickly transitioned away from placer mining districts to large-scale hard-rock operations, there were perhaps even more job opportunities in these areas for the Chinese. These largely male isolated hard-rock mining camps required the services afforded by Chinese labor,

Table 15. Montana employment of Chinese men, 1900 Federal Census

| Occupation | # of Chinese Employed |
|---|---|
| Laundry | 410 |
| Cook | 339 |
| Gardener or Farmer | 208 |
| Day or General Laborer | 182 |
| Mining | 126 |
| Merchant or Store Keeper | 110 |
| Restaurant | 80 |
| RR Laborer | 63 |
| Servant | 49 |
| Restaurant Waiter | 17 |
| Unknown | 13 |
| Patient | 12 |
| Tailor | 11 |
| Doctor | 9 |
| Dishwasher | 8 |
| Barber | 5 |
| Salesman | 5 |
| At Home | 3 |
| At School | 3 |
| Janitor | 3 |
| Store Clerk | 3 |
| Shoe Maker | 2 |
| Prisoner | 2 |
| Hotel Keeper | 2 |
| Landlord | 2 |
| Herder or Rancher | 2 |
| Hauling Goods | 2 |
| Baker | 1 |
| Gambler | 1 |
| Housekeeper | 1 |
| Wood Chopper | 1 |
| Restaurant Bookkeeper | 1 |
| Scavenger | 1 |
| Capitalist | 1 |
| Interpreter | 1 |
| Ice Cream Peddler | 1 |
| Woolen Mills Laborer | 1 |
| Teamster | 1 |
| Lodging House Operator | 1 |
| Real Estate | 1 |

Table 16. Montana employment of Chinese women, 1900 Federal Census

| Occupation | # of Chinese Employed |
|---|---|
| Wife/Daughter | 14 |
| Unknown | 8 |
| Prostitute/Sporting Women | 7 |
| Laundry | 3 |
| Keeping House | 1 |
| Restaurant | 1 |
| At Private School | 1 |
| Servant | 1 |

including laundry work, cooking, and gardening. Even with the restrictive laundry taxes levied on Chinese operations, they flourished and were ubiquitous across Montana (Jung 2008:79–80).

Chinese restaurants were also common across the Montana frontier, and sometimes dominated the local dining choices of communities, as seen in Big Timber in the 1880s–1890s (Moschelle 2009:19). While laundries and restaurants were the most common Chinese businesses, smaller numbers engaged in gardening, medical care, tailoring, servant/cook employment for private homes and businesses, and even stock raising. Chinese-owned general goods stores were discussed in chapter 1 in relation to German Gulch and Deer Lodge.

Rose Hum Lee (1949:51) developed a model of Chinese employment succession in Butte during the late 1800s and early 1900s, basing these changes on four factors: technological changes, discriminatory legislation and boycott actions, demographic changes, and finally intra-Chinese business competition. Lee (1949:50) notes that the Chinese in Butte, Montana, formed an occupational structure that complemented that of the larger non-Chinese society by undertaking roles and tasks required but not fulfilled by others. From the 1870s through the early 1910s the non-mining business niches filled by Chinese laborers and entrepreneurs were profitable, and opportunity abounded.

In her survey of Butte in 1890–1940, Lee (1949:51–55) created three Chinese business models that are relevant even in modern discussions of Chinese employment in Montana. The first type of business model was Chinese goods and services catering to an American clientele, which includes laundries, tailor shops, curio dealers, noodle parlors, and herb doctors (Lee 1949:51–52). A second type of Chinese business in Butte

CHINESE BUSINESS ESTABLISHMENTS IN BUTTE, MONTANA, FROM 1890 TO 1945

| YEAR | CHINESE GOODS AND SERVICES AMERICAN CLIENTELE | | | | | AMERICAN GOODS, CHINESE SERVICES, AMERICAN CLIENTELE | | | CHINESE GOODS AND SERVICES CHINESE CLIENTELE | | TOTAL |
|---|---|---|---|---|---|---|---|---|---|---|---|
| | Curio Shops | Noodle Parlors | Tailor-Shops | Laundries | Herb Doctors | Cafes, Restaurants | Truck Farms, Truck Vendors | Rooming-Houses | Merchandise Stores | Restaurants | |
| 1890 | ...... | ...... | 1 | 4 | 1 | 1 | ...... | ...... | 4 | 2 | 13† |
| 1895 | 3 | 1 | 3 | 18 | 3 | 2 | 2 | ...... | 5 | 3 | 40 |
| 1900 | 3 | 3 | 5 | 31 | 3 | 4 | 8 | ...... | 8 | 2 | 67 |
| 1905 | 3 | 7 | 8 | 31 | 5 | 4 | 2 | ...... | 7 | 2 | 69 |
| 1910 | 3 | 7 | 11 | 20 | 7 | 3 | 4 | 3 | 7 | 1 | 66 |
| 1915 | 1 | 6 | 7 | 21 | 6 | 3 | 5 | 4 | 7 | 1 | 60 |
| 1920 | 2 | 4 | 5 | 20 | 7 | 3 | 3 | 4 | 7 | 0 | 55 |
| 1925 | 2 | 4 | 2 | 11 | 7 | 3 | 2 | 5 | 6 | ...... | 42 |
| 1930 | 2 | 4 | 1 | 9 | 7 | 3 | 3 | 5 | 6 | ...... | 40 |
| 1935 | 1 | 4 | 1 | 8 | 4 | 3 | 2 | 5 | 5 | ...... | 32 |
| 1940 | 1 | 4 | ...... | 6 | 4 | 3 | ...... | 3 | 3 | ...... | 23 |
| 1945 | 1 | 3 | ...... | 3 | 2 | 3 | ...... | 2 | 1 | ...... | 14 |

* Sources: *City Directory, Butte, Montana* (1890–1942); *Silver Bow County Tax List* (1889–1945).
† Data incomplete.

Figure 12: Chinese businesses in Butte, 1890–1945. (From Lee 1949:53.)

sold American goods primarily to European Americans but had Chinese service workers, and these included restaurants, farms, and rooming/boarding houses (Lee 1949:54). The final type of Chinese business in Butte was owned and operated by the Chinese for a Chinese clientele and included traditional merchandise stores and restaurants (Lee 1949:55). These categories are relevant in today's study of the Chinese in Montana and beyond, though James and colleagues' (1994) study of Chinese laundries in Virginia City suggests that this type of business should be placed in both the first and third of Lee's categories.

## *Chinese Laundries in Montana*

A visitor to Montana Territory in the late 1880s through the first decade of the 1900s would have been hard-pressed to find a settlement of any notable size that did not have at least one Chinese laundry or restaurant. By the 1870 federal census there were Chinese laundries in many Montana communities, including at least one in Fort Benton (Robison 2009b). In 1880 there were 174 Chinese engaged in Montana's laundry business, with that number nearly tripling by 1900 to 410. Figure 12 illustrates the sheer number of Chinese laundries in operation in Butte, with 4 recorded in 1890, 18 in 1895, and rising to over 30 for the first decade of the 20th century. In comparison, a perusal of Helena's 1890 Sanborn Fire Insurance

Company map located 17 Chinese laundries, but these dropped to only a handful by the early 1900s. Lee (1949:53) states that laundry businesses were one of the simplest ways in which a Chinese individual could shift from being a working-class laborer to being an entrepreneur, as a laundry required little capital or excess labor and sustained a steady profit.

In some cases local non-Chinese residents objected to the presence of Chinese laundries over perceived health issues. For instance, in Fort Benton the construction of a new Chinese laundry elicited an editorial claiming that Chinese had created a health hazard by digging a cesspool for laundry and excrement run-off (*Fort Benton River Press* 1883:2). In contrast, most Chinese laundries left a positive and indelible mark in the memory of some Montanans, including the story of Quong Lun Kee of Dillon. Children of the locally prominent Orr family in Dillon remember receiving presents wrapped in their clean laundry when Quong Lun Kee returned it to the ranch. The standard and near-daily gifts included nuts, lily bulbs, incense, and firecrackers, with silver dollars and embroidered Chinese silk on holidays (Beaverhead County History Book Association 1990[2]:153, hereafter BCHBA).

According to James and colleagues (1994:168), there were two basic business models for the Chinese laundries of Virginia City, Nevada, and these also apply to Montana. Elderly Chinese men catering only to the Chinese community operated the first type of laundry business. In contrast, the second economic model was younger Chinese men who catered to the entire local community, regardless of ethnic affiliation. James and co-workers (1994:168) note that the elderly Chinese men merely operated a laundry as a means of housing and security, while the younger males sought to have a broad clientele and accumulate substantial wealth. Thus the first type of laundry was focused on a type of subsistence economy, while the second employed wage-earning labor and sought profit for material and social gain. As seen by the laundry tax and boycotts discussed earlier, it appears that the second type of Chinese laundry came under direct attack by the Montana community.

Laundries catering to the Chinese community were generally located in or adjacent to the Chinese part of town, while the second type were operated closer to non-Chinese domestic populations or boarding houses and hotels (James et al. 1994:169). This physical patterning is illustrated in Missoula, Montana, where there were several laundries in the Chinese district of that community, with one established on the far end of downtown and surrounded by exclusively non-Chinese populations and businesses. In addition the Chinese laundry apparently made use of the water canal (or race) of the Missoula Mills, which flowed directly behind the structure. The laundry operating in Missoula began in the late 1880s and was sold by 1900 to European American entrepreneurs. Given the loss of the 1890 census, it is impossible to determine the age of the laundry operators for this location, though they likely fit James and colleagues' (1994:168–169) model. According to the 1900 census, all 12 Chinese laundry employees and operators in Missoula fell below the age of 55, even those within the Chinese district on West Front Street.

There are numerous examples of historically known Chinese laundries in dozens of Montana communities from the 1860s through the mid-1900s. Only three, however, have been the subject of systematic archaeological excavation—one at Big Timber and two in Marysville. The excavated Marysville Chinese laundries were constructed side by side near the main road into town and operated from the mid-1880s through the early 1900s. Big Timber's Chinese laundry was constructed sometime after completion of the Northern Pacific Railroad and was not abandoned until the 1940s. Interestingly, the Big Timber Chinese establishment switched periodically between being a laundry and being a restaurant during those six decades, much of that time owned by Tom Kue (further discussed later in this chapter).

The Big Timber Chinese laundry excavation, occurring between 2007 and 2008, resulted in a professional paper for a master's degree by Justin Moschelle (2009) of the University of Montana. Big Timber formed in the early 1880s as a stop on the newly finished NPRR and quickly grew into one of the leading wool-producing hubs in the United States (Moschelle 2009:15). By 1900 the Chinese population of Big Timber reached 17, with most residents engaging in cooking or laundry work. Because the same Chinese-owned structure switched periodically from laundry to restaurant from the 1880s through 1940s, Moschelle's (2009) work serves as a reminder of Chinese practices in Montana. In turn, such dynamic business changes should be considered in urban archaeological contexts.

By the late 1930s Chinese laundries were an endangered business operation in Montana, though a few struggled through the 1940s. Boycotts and taxes since the mid-1860s had failed to destroy the Chinese laundry economic sphere, but in the mid-1900s a shift to steam laundries and home appliances signaled the end to one of Montana's most salient Chinese businesses. By the 1930 census only 74 Chinese worked in the laundry industry, a decline from 430 recorded in 1900.

### *Chinese Restaurants*

The growth and operation of Chinese restaurants in Montana follow a similar pattern to that of laundries. Chinese-owned restaurants spread throughout Montana following the major mining booms of the mid-1860s, though they became far more prevalent during the latter parts of the 1880s and into the early 1910s. These restaurants fell into three basic categories (following Lee 1949:51–55) determined by the types of food served. The first type of Chinese restaurant served traditional food and catered specifically to the Chinese community. A second type of restaurant served American and Americanized Chinese food to a broader clientele with the target consumer as non-Chinese. The third type served a broad variety of food that catered to both consumer groups; most modern Chinese restaurants in the United States fall into this type of category. In some communities, as in Big Timber (Moschelle 2009:15), the Chinese dominated the restaurant industry.

Only one Chinese individual in the 1870 federal census owned a restaurant in Montana, with this tally increasing to six according to the 1880 census. While these numbers likely underestimated the number of Chinese owning restaurants, they are an indication of the focus on mining and laundry operations during the first two decades of Montana Territory. Throughout the 1870s and 1880s, census records document several hundred Chinese cooks, most of whom found employment in non-Chinese establishments and private homes as well as restaurants owned by fellow Chinese residents. By 1900 the number of Chinese owning or working in restaurants increased to 80. The Butte boycott of the 1890s attempted to bankrupt Chinese businesses like laundries and restaurants, but also included any business that employed Chinese laborers, such as American-owned cafes.

According to Moses and Whitmore (1987:253), the Chinese in the United States exhibited a "cultural and ethnic cohesion to the point of ethnocentricity among even the lowest citizens of the 'Middle Kingdom,'" which "was not unlike behavior of first generation ethnic immigrants in the United States today." Many Chinese in the United States used traditional foodways as a means to reinforce their identity and to regulate their activities with non-Chinese practices (Moses and Whitmore 1987:254).

Figure 13: Small celadon or "wintergreen" style Overseas Chinese vessel discovered at Big Timber Chinatown in 2007. (Photo by author, 2007.)

In this way preparation and consumption of food helped to strengthen social and cultural bonds between different classes of the Chinese immigrant population (Moses and Whitmore 1987:258). Other scholars note important trends in Overseas Chinese foodways, including the propensity for pork over beef (Diehl and Thiel 1998:32), and the use of traditional butchering techniques (Longenecker and Stapp 1993:118), though not in all cases (Gust 1993:208). The most important conclusion of Diehl and Thiel's (1998:31–32) study was a determination that the Chinese gardener population of Tucson adapted to local conditions and prepared non-traditional foods in traditional ways.

There is little historical information on the interior workings of Chinese-owned restaurants in Montana, although archaeological work at Big Timber filled in some major gaps. As noted earlier, the Chinese business excavated at Big Timber by the University of Montana in 2007 and 2008 switched periodically between serving as a restaurant and laundry over the course of five decades. The artifact distribution at the site seems to suggest that the building served as a restaurant for a longer period than as a laundry, though the variable depositional processes of those businesses might skew the results (Figure 13). Of the 35,000 artifacts recovered from the site, crews recorded 2,569 faunal remains with a remarkable diversity. Faunal remains indicate that the Chinese were exploiting a wide variety of domestic animals and wild game for their restaurant and personal consumption. Species represented include cow, pig, goat/sheep, cat, rat, turtle, chicken, and grouse. Notably, crews failed to recover even a single fragment of fish remains, suggesting that perhaps this food source was underutilized or did not preserve well in the ground.

A unique finding at Big Timber was the recovery of 385 fragments of turtle shell (plastron and carapace) that likely derive from a locally procured species of painted or box turtle. Turtle was a common dish in traditional Chinese cuisine, in both the homeland and diasporic communities, and was usually boiled into a soup. Benn (2002:128) states that turtle soup was a gourmet dish in traditional Chinese society and might have reflected a higher-priced dish at the Big Timber establishment. Turtle might have been an exotic addition to the menu of the Chinese restaurant in Big Timber or might have been eaten solely by the proprietors and employees.

Other archaeological materials at the Chinese building at Big Timber suggest that the proprietors catered to a broad clientele. Excavators recovered a broad variety of condiment-related materials, including a complete 1883 Heinz Ketchup bottle, Lea & Perrins sauce stoppers, and possible mustard and pickle vessels. In addition there were a significant number of alcohol bottles, including domestic and imported whiskey, beer, champagne, wine, and several Gordon's Dry Gin bottles imported from England. In keeping with Lee's (1949) model of Chinese businesses at Butte, it appears that the Chinese restaurant in Big Timber catered to a diverse population but likely served mostly American or Americanized Chinese food to the local non-Chinese population.

Unlike the laundries, the Chinese restaurants never completely left the Montana landscape, though their number dipped considerably statewide after 1900. Today most Chinese restaurants in Montana focus on appealing to the dominant non-Chinese host community. In most cases Montana Chinese restaurants serve Americanized Chinese food, though there are a few (e.g., in Butte, Bozeman, Livingston, and Missoula) that provide authentic Cantonese cuisine.

### *Chinese Gardens*

Many Chinese immigrants who left China in the late 19th century came from an agricultural background, and these skills found use in isolated mining communities in Australia, New Zealand, Canada, and the United States (Valentine 2002:44; Dirlick and Yeung 2001:81, 126). On the Montana mining frontier, fresh produce was a welcome rarity to the isolated camps. During the initial gold strikes of the mid-1860s there were only a handful of farms in Montana Territory, with Major Owen in the Bitterroot Valley doing a brisk business freighting produce to the boomtowns. Even as Montana's population increased and farms sprang up, there was always a produce deficiency, and the Chinese capitalized on this shortcoming. As best stated by Zhu (1997:114), "a well-cultivated garden often produced more gold than a mediocre placer field." Scholars of the Overseas Chinese have only touched briefly on the history and archaeology of Chinese market or truck gardens around the world (Stanin 2004; McGowan 2005). In Australia McGowan (2005) surveyed a series of Chinese gardens in New South Wales and noted that they played an important role in the economic development of Australia throughout the late 1800s and early 1900s. In contrast to Lee's (1949:50) determination that these gardens focused on an American clientele, it is more likely that they provided produce to the entire community. In New Zealand Chinese merchants owned the larger gardens in order to supply goods to their stores that serviced a broad consumer base (Ritchie 1986:640); this pattern likely existed in Montana, though there is little historical documentation.

Fee (1991; 1993) studied a series of terraced Chinese gardens along the Snake River in Idaho and found that they produced a wide variety of crops. Both McGowan and Fee's garden studies noted how the Chinese were able to produce substantial crops in an arid environment, which is likely a testament to centuries of water conservation in agricultural China. In Australia the Chinese market gardens collapsed in the post–World War II period (Williams 1999:26), while the abandonment of this lifestyle occurred in Montana largely during the 1910s–1920s. As noted by Ritchie (1986:640), it was far more likely for Chinese in the Australian gold fields to maintain gardens than for their non-Chinese counterparts, especially in the earliest days of mineral exploitation.

The 1870 Montana census lists no Chinese gardeners, though by 1880 there were 20 employed throughout the state, clustering in Butte, Helena, Virginia City, Ophir Gulch, and the Prickly Pear Valley. By 1900, with the relative collapse of the placer mining opportunities for the Chinese, the number of gardeners increased substantially to 208, becoming the third-largest employment category behind laundry work and cooking. Helena had the largest share of Chinese gardeners in the state in 1900, though Butte had numerous operations south and west of the city. Other communities that had Chinese gardeners included Missoula, Dillon, Kalispell, Billings, Fort Benton, Boulder, Anaconda, and Bozeman.

Chinese gardens were an important part of Montana's 19th-century communities, as the predominantly male populace did not often engage in subsistence agriculture and gardening. In the first few decades of Montana's settlement history, Chinese gardeners cornered the market on fresh produce, creating a steady livelihood throughout the spring and summer months. In Fort Benton, Chinese gardeners harvested the first crops of the season in 1883 and sold onions, spinach, and celery at a high price (*Benton Record* 1883:2). Quong Lee, a Deer Lodge merchant, operated a small garden, and a reporter for the *New Northwest* acquired a number of cucumbers from him in 1881 (*New Northwest* 1881b:3). By 1885 there were apparently a few dedicated Chinese gardeners around Deer Lodge who brought radishes, lettuce, and asparagus to town for above-market price due to limited supplies (*New Northwest* 1885a:3). Dillon had a series of large Chinese gardens on the north end of town (BCHBA 1990[2]:974) that operated between 1880 and the 1920s (BCHBA 1990[1]:9). Whiskey Jim, a Chinese gardener and former miner, lived in Dillon until his death in the 1930s and sold produce to "housewives, grocers, restaurants and chop houses" (BCHBA 1990[1]:10).

Helena's downtown, near the mouth of Last Chance Gulch, had several large Chinese gardens beginning in the late 1860s through late 1890s. By 1882 there was a shift in the type of businesses operated in Last Chance Gulch, with one acre previously known as "Chinese Gardens" sold to non-Chinese businessmen for nearly $20,000 (*Helena Independent* 1882i:3). Sanborn Fire Insurance maps from 1888 show several Chinese gardens along West Main, Cutler, and Water streets, but by 1930 there were none illustrated. According to the 1930 census, some of the Helena Chinese gardens moved to the north end of town, where space was more plentiful. Even these gardens failed to survive the 1930s, and yet another Chinese business venture disappeared from the Helena landscape.

Missoula had two areas of Chinese gardens during the early part of the 1900s. The eastern Chinese gardens clustered around the opening to Hellgate Canyon, between the Clark Fork River and Mount Jumbo (*Anaconda Standard* 1900c:10, 1900d:12). There was a second, and seemingly larger, cluster of Chinese gardens west of the downtown at the same time, apparently located west of modern-day Russell Street (*Anaconda Standard* 1901a:12). A series of violent robberies occurred at Missoula's western gardens in 1901, with several thousand dollars stolen from the Chinese gardeners and several non-Chinese arrested for the crimes (*Anaconda Standard* 1901b:12, 1901c:10, 1901d:12).

Overall the Chinese gardeners of Montana have received little formal attention from historians until recently, although urban development has destroyed most of the original plots. Urban renewal and redevelopment destroyed the remaining segments of Helena's downtown gardens in the 1900s, while housing subdivisions now cover the rural plots north of downtown. The only remaining areas of potential intact Chinese gardens are at Glendale, Ophir Gulch, and Marysville. Hall and colleagues (2006:42) note that locals suggest there are Chinese gardens in the northwestern part of Marysville, and surveyors located but did not excavate what appears to be a raised garden bed at the rear of the laundries in that community. Local history in Plains, Montana, suggested that Chinese gardeners constructed rock terraces in Poacher Gulch during the 1880s and 1890s, though archaeological investigations shed serious doubt on those claims (Merritt 2006, 2009b).

The best specific historical information on Chinese gardens and gardening techniques comes from a 1906 *Anaconda Standard* story on Butte's Chinatown, which was included in full in a recent book (Dean 2009). During the early years of Butte's Chinese garden industry, their plots centered on the flat below downtown and near the railroad tracks. Due to shifting property values, the Chinese moved their operations to the confluence of Basin and Ninemile creeks, to the southeast of downtown (Dean 2009:351). Workers at these gardens averaged about eight per operation, though one owner of a 15-acre plot, Ye Dee, stated that finding enough employees was difficult and he only had seven (Dean 2009:349). A 10-acre garden generally produced about $8,000 in crops per year, with $500 of that going toward rent and several thousand dollars going to pay for labor (Dean 2009:347, 350). Due to Butte's high altitude, the gardening season lasts a total of five months, though three are spent in preparation and two represent the actual growing season (Dean 2009:350). Ye Dee produced a variety of market crops, including turnips, carrots, beets, onions, lettuce, peas, corn, and only a small batch of potatoes, grown in open fields and in rows of covered hothouses (Dean 2009:347–348, 351). The covered hothouses were apparently an adaptation to the unpredictable and short growing season of Montana, as Chinese market gardens in Australia did not employ this type of technique (Stanin 2004). Unlike in Idaho (Zhu 1997:113), it appears that the Butte Chinese did not grow a substantial amount of Chinese cabbage.

By 1930 the number of Chinese gardeners dropped to 31, and these were mostly located in Butte and Helena. Small-scale Chinese gardeners were simply incapable of keeping up with the mechanized farm and the low-cost importation of produce from other areas of Montana and North America. In addition, the arrival and succession of Korean and other ethnic groups into the Butte gardening business also forced the Chinese out of operation (Lee 1949:58).

### *Chinese Cooks and Servants*

Domestic labor served as another avenue for Chinese to find gainful employment throughout the 19th and 20th centuries in Montana. Cooks and servant positions were by far the second-largest pools of Chinese labor after 1900 but accounted for third place in the 1870s and 1880s. In 1870 there were 37 cooks and 19 servants, but these numbers increased significantly, to 238 and 26 respectively, in 1880. By 1900 employment as cooks accounted for 20% of the entire Chinese workforce at 338, with servant positions rising to 50. Cooks became the largest occupation category for Chinese in Montana by 1930 at 86, again accounting for 20% of all employment. Census takers recorded only one servant in 1930, which might be an indicator of the changing terminology for domestic labor and also of the impact of the economic downturn and repercussions of the earlier boycotts. Cooks discussed here are wage laborers and include those occupying positions in hotels, railroad camps, sawmills, hard-rock mine camps, and both Chinese and non-Chinese restaurants. This does not include Chinese individuals who owned or operated restaurants, as they would be listed in the census as proprietors or owners.

Chinese cooks and servants worked in a much broader geographic swath of Montana than any other occupational groups. Throughout the last four decades of the 19th century Chinese cooks and servants spread throughout the vast eastern prairie, sometimes representing the only Chinese person within a several-hundred-mile radius. In many of the rural farming and ranching communities a Chinese cook was a common sight and actually was a sought after commodity to decrease the workload for the female household members. The large Orr family of Dillon employed several Chinese cooks from the 1880s through 1920s who became an unofficial part of the family (BCHBA 1990[1]:153).

Chinese cooks and servants worked in many of Montana's military posts, either working for individual soldiers or providing services to the entire garrison. In 1880 Fort Shaw had five Chinese servants employed by officers, including Colonel Brooke, Sergeant Greenleaf, and Lieutenants Hennay, Jones, and Hale. That same year there were three Chinese men, two laundry workers, and a cook employed at Fort Logan, four laundrymen at Fort Keogh in Custer County, and three servants at Fort Missoula. Archaeological excavations at Fort Missoula discovered several examples of Chinese celadon, though supervisors never completed a final report detailing the discoveries. A list of interments at the Fort Shaw Post Cemetery in 1881 notes the burial of Joe Chinaman, a servant to Captain Morely, 3rd infantry (United States Army 1881:251). Additional research discovered that Joe Chinaman was likely Joe Wanman, a Chinese servant in the officers' quarters who committed suicide in 1881 by pistol (*Helena Independent* 1881g:3). Lee Yee, a Chinese servant, died at Fort Custer in 1891 and was interred at the post cemetery. The number of Chinese on Montana's military posts declined precipitously in the first decade of the 1900s, with Jim Lee, a laundryman in Fort Harrison, recorded as the only one in 1900.

Some former Chinese cooks and servants rose to prominence in Montana, including Billy Kee of Lombard. According to some accounts, Billy Kee was the servant and cook for United States Senator Thomas Carter and family in the 1870s and 1890s. When the NPRR sponsored the construction of a spur to run from the main line northward to Lewistown and other central locations of Montana in the 1890s (Stoner 2007:33), Lombard was established as a hub. Kee acquired the hotel through some assistance from his former employer, Senator Carter, ran the operation until around 1910, and departed Montana for China in 1914 (Stoner 2007:34). During his stay at Lombard, Kee successfully brought his wife and children from China, and a famous photograph of the family hangs in the Montana Historical Society today.

When Billy Kee and family left Montana for China, it was shortly after the revolution in that country, and rumors circulated of Kee's execution. Several months later Kee was found alive in Japan, not even aware of his reported execution (Stoner 2007:38–39). Perhaps one of the most interesting aspects to Billy Kee's story was the supposed use of the fortune he amassed in Montana to help fund the revolutionary overthrow of the Qing

Dynasty. As Stoner (2007:33–34) notes, Billy's life was interesting, though the exact details are shrouded in myth, hearsay, and likely exaggeration. It is unclear how a former servant like Billy Kee became the wealthiest Chinese in Montana by operating a hotel for several years, though that is perhaps a reason for the popularity of his life's story.

Billy Kee's life is an exception to the more general pattern of long-term wage labor of the Chinese cook or servant in Montana history. As noted in chapter 1, as early as the 1860s Major John Owen was employing Chinese cooks at Fort Owen to provide meals for his workers and guests (Dunbar and Philips 1927[2]:58, 67–68). Chinese cooks and servants served notable Montana families, including in the homes of Charles Conrad, founder of Kalispell; George Beal, famous Butte doctor and mine owner; William Wickes, founder of the town of Wickes; A. B. Hammond, early merchant in Missoula; Samuel Hauser, powerful Helena banker and mine developer; and Judge Knowles of Deer Lodge.

Judge Knowles passed judgment on several Chinese-related court cases, including on the Montana Alien Law and on the murder of two Chinese men, one by John Collins and the other by John Martell from (all described in chapter 1). In all cases Knowles appears to have possessed pro-Chinese sympathies, including the conviction of John Collins for murder, setting a precedent in allowing Chinese testimony against John Martell and overturning the Alien Law. Perhaps Knowles's use of Chinese labor in his household shaped his sympathies toward this population. It is clear that the presence of Chinese cooks and servants improved the quality of life for hundreds of Montana families and provided these immigrants with a steady source of income and job security. These jobs, located on rural farms and ranches, led to the situation of severe isolation for the Chinese cooks and servants.

### Other Notable Chinese Employment

It is clear that laundries, restaurants, gardens, and domestic labor provided the bulk of Chinese employment opportunities in Montana after the collapse of placer mining in the late 1880s. Broad patterns can sometimes cloud the personal and significant individual stories of Chinese who forged their own niche in Montana history. The various censuses between 1880 and 1940 provide some information on other employment associated with the Chinese, and many break the generally constructed narratives of Chinese as miners, launderers, or gardeners. For instance, the 1880 census listed a wide variety of jobs, including priests, doctors, druggists, and cheese makers—all occupations defying the long-held stereotypes of Chinese employment in the United States.

Chinese physicians were common in larger communities on the Montana frontier and also some rural communities like Louiseville (see chapter 1). From the 1870s through the 1930s there were Chinese physicians in Butte, Billings, Deer Lodge, Helena, Livingston, Louiseville, and Virginia City. In Helena prominent Chinese doctors included Dr. Fook On Hong Tong and Yeng Too. Willie Toy was a prominent Helena and Anaconda-based Chinese physician and herbologist who trained in San Francisco using money earned from a string of restaurants in the early 1900s (*Helena Independent* 1929b:4). Huie Pock of Butte was by far the most prominent Chinese doctor in Montana during the 1800s and early 1900s. He arrived in Butte in 1889 with his young wife and stumbled almost immediately into the Butte Union Boycott, which deeply affected his medical and merchant businesses (Dirlick and Yeung 2001:400–401). Dr. Pock is also well-known in Montana history for curing the ulcers of "Copper King" William A. Clark's daughter and for saving countless lives in Butte during the international influenza outbreak 1918 (Dary 2008:194). Pock and his wife drew up a will specifying that their estate would return their bodies to China after their deaths. Ultimately the European American mortician kept the bodies in his funeral home for over 30 years, one body sealed in a casket and the other treated with lubricants. Finally, after 30 years in the funeral home the bodies of Huie Pock and wife were buried in Mount Moriah Cemetery, as there was no money left in the estate to return the remains to China (Ciabattari 1976:16).

Among the most distinctive stories of any Chinese experiences in Montana are those involving Tommy Haw and George Washington Taylor. Their stories began rather similarly, both being born in China in the middle 1800s and brought to the United States by immigrating relatives or parents. George Taylor was born in Hong Kong in 1860 and was brought to Helena in 1873 by an uncle (Hilden 1942); in another version his mother died, leaving him an orphan in Helena (*Mineral Independent* 1930:4). In the most likely scenario, recounted by George Taylor himself, his uncle set George up as a servant for a Helena lawyer and then returned to China. In contrast, it is unclear what occurred in Tommy Haw's family, but some stories suggest that his father died in the gold fields of California, leaving him an orphan at age four (BCHBA 1990[1]:433).

The exact year of birth for Tommy is rather unclear, as the 1880 census records indicate he was born in 1850, while his obituary states he was born in 1846 (*Dillon Tribune* 1913:2). Rancher Tom Orr found Tommy wandering the streets of San Francisco and adopted him around 1850 or 1854 (depending on his actual date of birth). Another version of the adoption of Tommy suggests that his parents died in an Indian raid on a Chinese mining camp near Yreka, California, and Orr found him as the last survivor (BCHBA 1990[1]:431). In 1865 Tommy came to Dillon, Montana, with a herd of Orr's cattle and worked with the family on a large ranch near that town for years to follow (BCHBA 1990[1]:433). Upon reaching Montana, one of Haw's first actions was to cut his long braided queue, which, in hopes of ensuring acceptance back into the Chinese Community, Tom Orr had forced him to keep up until this point. Sometime in the 1870s Tommy Haw went into the stock-raising business for himself with assistance of the Orr family. After selling a herd of his cattle for $9,000, he used the money to purchase a sheep herd and lease a ranch (BCHBA 1990[1]:431).

By the 1880s Tommy was one of the wealthiest citizens of Beaverhead County, paying a substantial amount in taxes (*Dillon Tribune* 1883:1), and was well-regarded by locals and the newspapers (*New Northwest* 1883:3). In 1886 Johnny Grant, the famous founder of the Grant Kohrs Ranch, pulled into Tommy's ranch late at night. Haw's white employees took care of Grant's horses but neglected to give him dinner. Tommy was most displeased about this the next morning and provided Grant with a large meal. When Grant asked Tommy what the services would cost, Haw replied, "Ah! Nothing, nothing, when you see a Chinaman . . . in need, give him a meal" (Grant 1996:198).

In the late 1880s Haw was purchasing mining claims in the Pioneer Mountains, west of Dillon, including the Copper Queen Mine. Perusal of the Beaverhead County Courthouse records indicate that Tommy purchased several mining claims in the region during the late 1880s through early 1900s, along with additional ranch leases. A 1903 deed record indicates that a local Chinese resident, Tommy Haw, purchased the Blue-Eyed-Annie claim (24BE2186) and managed the work until 1909–1910. In 2008 researchers rediscovered the site as part of my dissertation fieldwork and recorded the remaining log cabins, adits, trash dumps, and underground workings (Merritt 2010).

It appears that 1903 might have been the peak of Tommy Haw's fortune, with newspaper reports claiming his estate and assets came to $200,000

(*Pittsburgh Press* 1903:3). Tommy's business sense did not appear to extend into the mining business, and by his death in 1913 his fortune had declined significantly. Suffering from a long liver-related illness, Tommy died on July 25, 1913, at the suspected age of 67 (*Dillon Tribune* 1913:1). After his death Haw's will was located, and he had left all his property and possessions to a Mrs. George Gray of Dillon, who cared for him during the time of his illness. In addition, Tommy's will stipulated that the revenue from the pending sale of the Copper Queen mine for $37,000 would also go to Mrs. George Gray and two other white women who cared for him during his illness (BCHBA 1990[1]:431). Due to his poor finances at the time of his death, Haw's estate lacked enough funds to purchase a headstone. Friends and family collectively purchased Haw's headstone, and he now rests within the Orr family plot in the Dillon Cemetery (BCHBA 1990[1]:431). In the few years before his death Tommy continuously petitioned to be naturalized as an American citizen, though he was never successful.

George Washington Taylor came to Montana from Hong Kong in 1873 and settled in Helena as a lawyer's servant (Hilden 1942). Shortly after his arrival in Helena either by death of his mother or by abandonment of his uncle, George, whose Chinese name is unknown, started learning to speak and write English. Sometime after his mother's death, Jesse F. Taylor of the Sands and Taylor Cattle Company (ST) adopted the Chinese boy and renamed him George Washington Taylor (*Mineral Independent* 1930:4); George Taylor stated in 1942 that he renamed himself, as "my imagination and fancy were drawn toward George Washington, the father of this country" (Hilden 1942). Through the 1870s George worked as a stockman and cowboy for the ST and had numerous interesting adventures as part of his life working the range in the Choteau and Cut Bank area of Montana.

The 1880 census lists George Taylor erroneously twice, once in Fort Benton and once on the Upper Teton River; however, both entries listed him as a 19-year-old cook. Around 1879 the Jesse Taylor gave George and his white brothers a herd of cattle to start their own ranching endeavors (Hilden 1942). In the 1880s George married a Swedish woman named Lena from Minnesota, and by 1900 they had two children. The 1900 census lists the Taylor family living in Dupuyer Township, in Teton County, and involved in farming. According to other sources George and family established a small stock ranch near Birch Creek (*Mineral Independent* 1930:4). According to several reports, the Taylor family established a steady operation in the Birch Creek area, and their home was one of the finest in the local country and a favorite stopping place for travelers (*Mineral Independent* 1930:4; Hilden 1942).

George and Lena Taylor, luckily, avoided the adverse effects of Montana's miscegenation law. In 1909 Montana's legislature passed the state's first miscegenation law barring intermarriages between "whites" and other ethnicities, such as African Americans, Chinese, or Japanese (State of Montana 1909:57–58). The language of Montana's miscegenation law prohibited or voided marriages performed after 1909, leaving George and Lena's 1880s or 1890s marriage legal until her death in the 1920s. Sadly, others in the African American, Chinese, and Japanese community felt the repressive effects of this racist law until 1953, when it was finally repealed by the legislative assembly (State of Montana 1953:4–5).

Sometime in the 1920s George Taylor's wife passed but not before building a sizable family legacy including seven children. The 1920 census lists George, Lena, and their six living children Albert, Margaret, George Jr., Helen, Andrew, and Daisy. Their second oldest son, Jesse, died in World War I, though Hilden (1942) incorrectly states that Albert was killed. In both 1900 and 1920 the Taylors were boarding the local schoolteacher, which is a sign of their complete integration into the local community. By 1930 Taylor left his ranch and moved to Cut Bank after Lena's death, and two of his sons engaged in sheep ranching on the Blackfeet Reservation while the third enlisted in the United States Navy (*Mineral Independent* 1930:4).

By the 1940s George had settled into a comfortable small home in Cut Bank, with his sons continuing their operations in the region and his daughters living in Alaska and San Francisco. During his interview for the *Cut Bank Pioneer*, Taylor reminisced about his life spent in Montana, including when Charles McDonald, a Teton County representative, put forth an amendment to allow Taylor's naturalization as an American citizen (Hilden 1942). Taylor's naturalization was celebrated as the first in Montana for a Chinese resident, an honor denied to Tommy Haw years earlier. Hilden (1942) concludes her story on George Taylor with poignant prose:

*The young Chinese boy who left Hong Kong when he was only 13 years old, now lives in a neat little house in the shaded yard behind the Cut Bank hotel. His house is full of mementos of a long life spent in Montana—pictures of his ranch and his family. He's like any other venerable gentleman in his declining years. He's an American.*

The last statement of Hilden (1942) is as much a celebration of George Taylor's life as a rapprochement of the maltreatment of the other Chinese in Montana who could have benefited from the opportunities provided him. George Taylor died in Cut Bank on May 15, 1945, at the age of 85, and lies buried there today.

## Sun Descending over the Horizon, 1900 to 1943

As changing circumstances shifted the economy in the first decades of the 20th century, Chinese opportunity continued to decline, leading directly to the abandonment of Montana for areas with higher potential for profit and job security (Lee 1949:58). The anti-Chinese immigration legislation effectively cut off the population in Montana, leading to further migrations out of the state. In 1900 there were 1,722 Chinese in Montana, but the number declined to 1,285 a decade later. The population collapse continued, with numbers falling to 872 in 1920, 491 in 1930, 258 in 1940, and finally bottoming out at 209 in 1950 (Tables 17 and 18). There was also a consolidation of Chinese population from rural to urban settings, as seen from the locations and numbers of places with Chinese residents (Figures 14, 15, 16, and 17). Some scholars (Swartout 1988; Zhu 1997) discuss the year 1900 as the end of the historical period of Chinese settlement in Montana. While the population peak of the Overseas Chinese had passed by 1900, those still living in Montana continued to make significant contributions to state history and strengthened their connections to China. With the repeal of the Chinese Exclusion Act in 1943, the history of Chinese in Montana changed tone, and for the first time since the mid-19th century Chinese people were allowed to emigrate to Montana, although on a quota system.

The laundry industry dominated the 1900 employment of Chinese, but by 1930 and 1940 cooks and "unknown" took the top spots (Tables 18 and 19). The unknown category likely represents the number of unemployed Chinese in Montana, which resulted from both the economic downturn from 1929 and also the advanced age of many enumerated individuals.

The average age of Chinese in Montana continued to increase throughout the 1800s and early 1900s. In 1870 the average age of a Chinese person in Montana was 30, and by 1880 that number rose to 35. In 1900 the average age of a Chinese individual was 42, with the oldest being 72 years old (Figure 18). The 1930 census shows that the average age remained steady at 42, with a median age of 43.5, but continued an upward trend to an

Table 17. Montana Chinese population by county, 1900–1930, Federal Censuses

| County | 1900 | 1910 | 1920 | 1930 |
|---|---|---|---|---|
| Beaverhead | 67 | 33 | 17 | 7 |
| Big Horn | 0 | 0 | 12 | 0 |
| Blaine | 0 | 0 | 4 | 6 |
| Broadwater | 14 | 2 | 6 | 0 |
| Carbon | 2 | 0 | 3 | 0 |
| Cascade | 3 | 6 | 5 | 2 |
| Choteau | 87 | 40 | 8 | 0 |
| Custer | 16 | 31 | 28 | 18 |
| Daniels | 0 | 0 | 0 | 5 |
| Dawson | 2 | 14 | 5 | 3 |
| Deer Lodge | 88 | 26 | 23 | 18 |
| Fallon | 0 | 0 | 8 | 4 |
| Fergus | 14 | 15 | 25 | 12 |
| Flathead | 58 | 61 | 37 | 1 |
| Gallatin | 54 | 62 | 14 | 3 |
| Glacier | 0 | 0 | 5 | 11 |
| Granite | 69 | 25 | 16 | 2 |
| Hill | 0 | 0 | 49 | 19 |
| Jefferson | 57 | 23 | 7 | 2 |
| Lake | 0 | 0 | 0 | 1 |
| Lewis and Clark | 330 | 328 | 135 | 87 |
| Lincoln | 0 | 5 | 3 | 8 |
| Madison | 77 | 10 | 3 | 1 |
| Meagher | 5 | 20 | 10 | 7 |
| Mineral | 9 | 0 | 1 | 1 |
| Missoula | 113 | 73 | 18 | 14 |
| Park | 42 | 38 | 15 | 11 |
| Phillips | 0 | 0 | 10 | 0 |
| Pondera | 0 | 0 | 2 | 2 |
| Powell | 0 | 14 | 11 | 20 |
| Prairie | 0 | 0 | 1 | 0 |
| Ravalli | 30 | 21 | 9 | 2 |
| Roosevelt | 0 | 0 | 7 | 3 |
| Rosebud | 0 | 4 | 1 | 0 |
| Sanders | 61 | 38 | 18 | 5 |
| Sheridan | 0 | 0 | 5 | 0 |
| Silver Bow | 389 | 319 | 262 | 155 |
| Stillwater | 0 | 0 | 2 | 0 |
| Sweet Grass | 18 | 14 | 17 | 8 |
| Teton | 17 | 3 | 2 | 1 |
| Toole | 0 | 0 | 2 | 1 |
| Valley | 0 | 10 | 2 | 2 |
| Wheatland | 0 | 0 | 2 | 0 |
| Yellowstone | 99 | 50 | 55 | 49 |
| (All other counties) | 0 | 0 | 7 | 0 |
| Totals | 1,722 | 1,285 | 872 | 491 |

Table 18. Montana employment of Chinese men, 1930 and 1940 Federal Census

| 1930 Census | | 1940 Census | |
|---|---|---|---|
| Occupation | # of Chinese Employed | Occupation | # of Chinese Employed |
| Cook | 86 | Unknown/None/Child | 49 |
| Unknown/None/Child | 86 | Cook | 42 |
| Laundry | 74 | Merchant/Store Keeper | 30 |
| Gardener/ Farmer | 31 | Restaurant Waiter | 23 |
| Prisoner | 30 | Laundry | 17 |
| Merchant/Store Keeper | 21 | Gardener/Farmer | 9 |
| Restaurant Waiter | 19 | Clerk | 7 |
| Restaurant | 17 | Inmate | 6 |
| Laborer | 9 | Laborer | 6 |
| RR Conductor | 7 | Herb Merchant | 4 |
| Hotel Keeper | 6 | Dishwasher | 4 |
| Dishwasher | 5 | Janitor | 4 |
| Doctor | 4 | Student | 3 |
| Ironers | 3 | Ironers | 2 |
| Tailor | 3 | Prospector | 1 |
| Cigar Salesman | 3 | Elevator Operator | 1 |
| RR Laborer | 2 | Doctor | 1 |
| Real Estate | 2 | | |
| Mining | 1 | | |
| Retired Cook | 1 | | |
| Servant | 1 | | |
| Groceries Bookkeeper | 1 | | |
| Court Interpreter | 1 | | |
| Janitor | 1 | | |
| Teacher | 1 | | |

average of 45.6 in 1940 (Figures 19 and 20). Substantially different from the 1900 census, the oldest Chinese person in Montana was a 92-year-old woman, with an additional 36 men older than 72 years old. An aging population typified the first half of 20th-century Chinese history in Montana, although there were a number of children recorded in the 1930 and 1940 censuses, most belonging to a dozen or so families in Butte and Helena. The presence of just a few families with children skew the population charts for 1930 and 1940, representing a second, and extremely small, demographic change.

Table 19. Montana employment of Chinese women, 1930 and 1940 Federal Census

| **1930 Census** | | **1940 Census** | |
|---|---|---|---|
| Occupation | # of Chinese Employed | Occupation | # of Chinese Employed |
| Unknown or Child | 50 | Unknown or Child | 24 |
| Stenographer | 2 | Waitress | 3 |
| Merchant | 1 | None | 1 |
| Laundry | 1 | Bookkeeper | 1 |
| Collection Agency | 1 | Housekeeper | 1 |
| Gardener | 1 | Stenographer | 1 |
| | | Seamstress | 1 |
| | | Landlady | 1 |
| | | Inmate | 1 |
| | | Housewife | 1 |
| | | Café Operator | 1 |

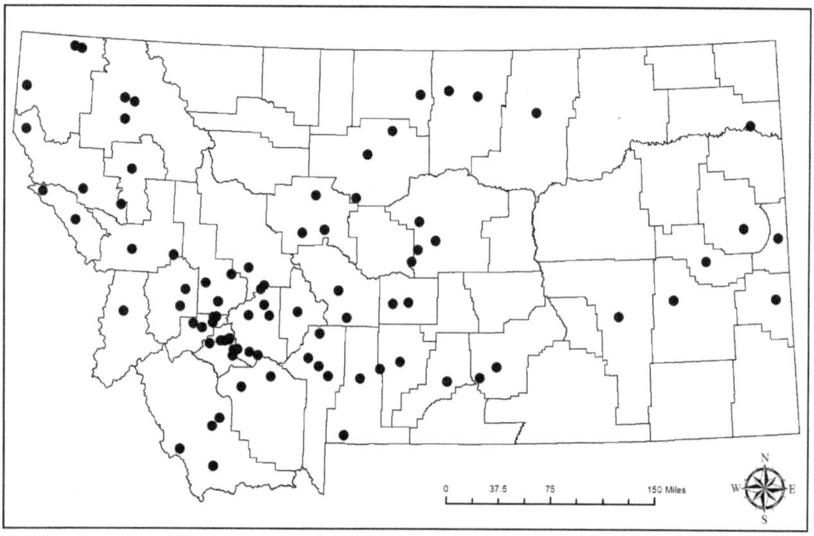


Figure 14: Chinese population centers in 1910. (Created by author.)

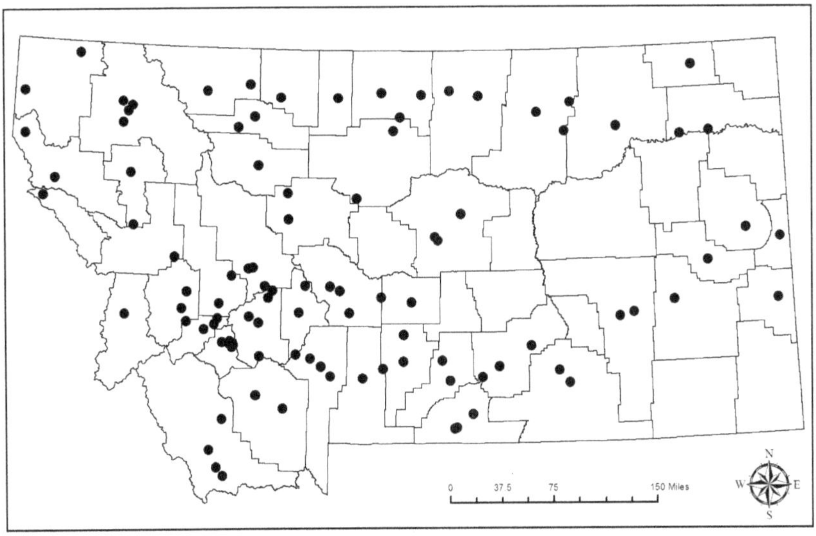


Figure 15: Chinese population centers in 1920. (Created by author.)

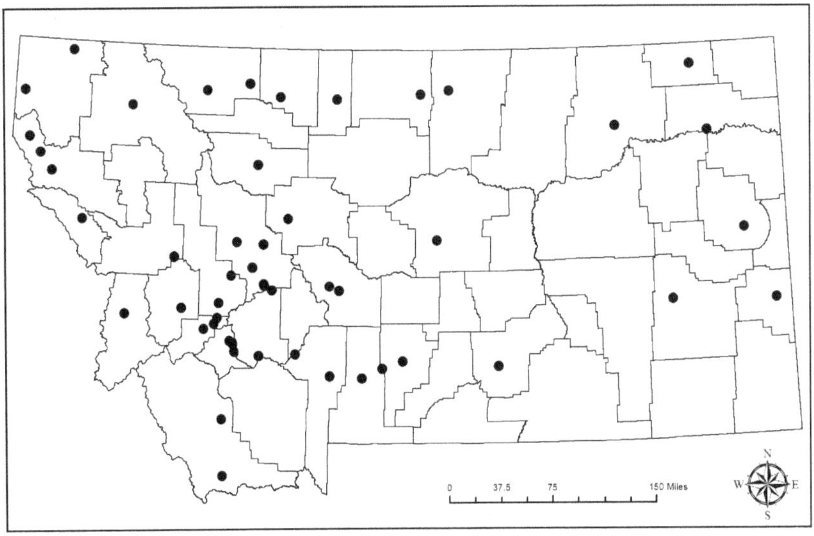


Figure 16: Chinese population centers in 1930. (Created by author.)

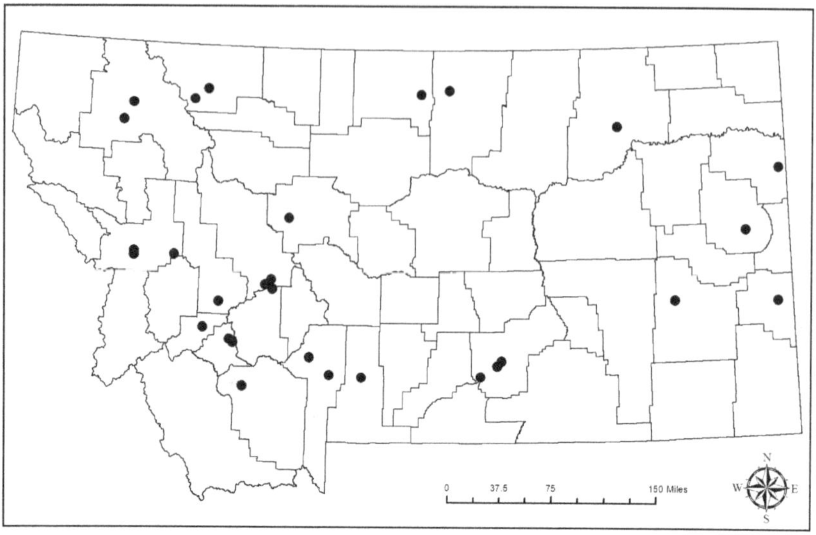


Figure 17: Chinese population centers in 1940. (Created by author.)

Between 1900 and 1940 the Chinese population of Montana consolidated into the larger urban and rural centers. In 1900 there were Chinese in 113 distinct enumeration districts, while in 1930 that number declined to only 51 and even further to only 28 in 1940 (Tables 20, 21, and 22). This consolidation process resulted from the overall Montana population of Overseas Chinese declining from both death and out-migration. Butte and Helena retained the top two positions for Chinese population in Montana during the entire first three decades of the 20th century. Billings retained a significant Chinese community from the 1890s through 1940, and by that latter year, the city's Chinese population ranked second in the state, behind Butte. The biggest drop in Chinese occurred in Missoula, with 109 in 1900 but only 10 enumerated in 1930 and 3 in 1940.

It is clear that the anti-Chinese agitation that formulated the Chinese Exclusion Act succeeded in its effort to cleanse the state of this ethnic group, though their spirit remained strong. Throughout the early 1900s some Chinese continued to celebrate their traditional rites and the New Year's ceremonies, while others, like the Wong family of Helena, attempted to assimilate by moving away from tradition (Mike Wong 2008, pers. comm.). The Wong family in particular contradicts the trend in historical archaeology of moving away from assimilation/acculturation discussions,

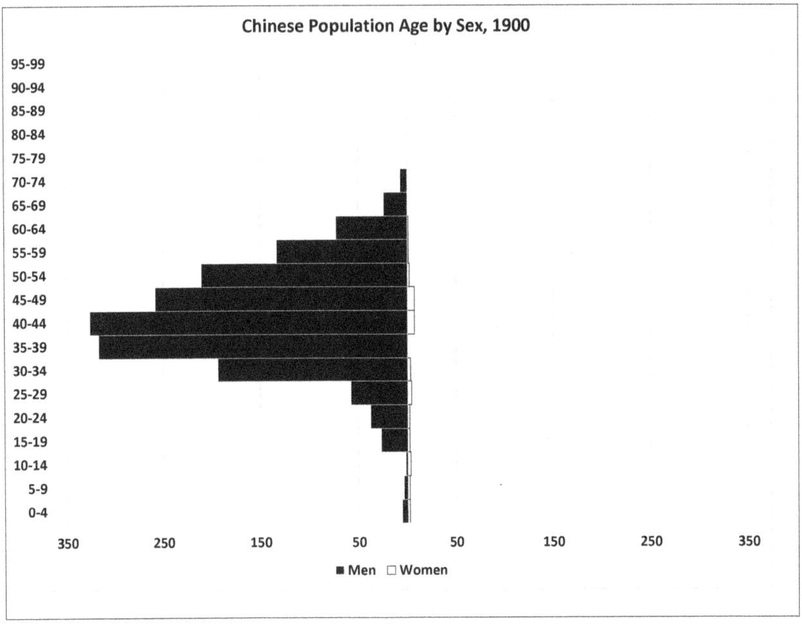


Figure 18: Population pyramid of Chinese in Montana, 1900. (Created by author.)

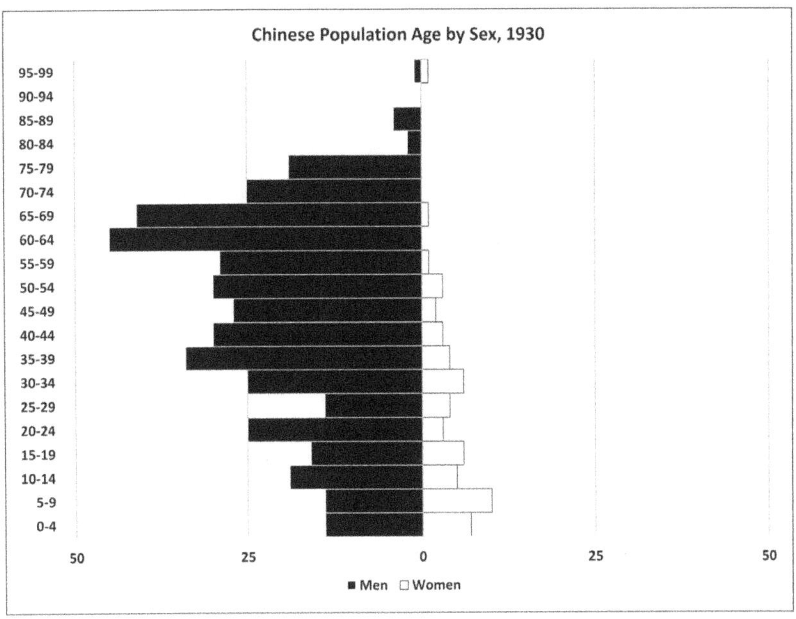


Figure 19: Population pyramid of Chinese in Montana, 1930. (Created by author.)

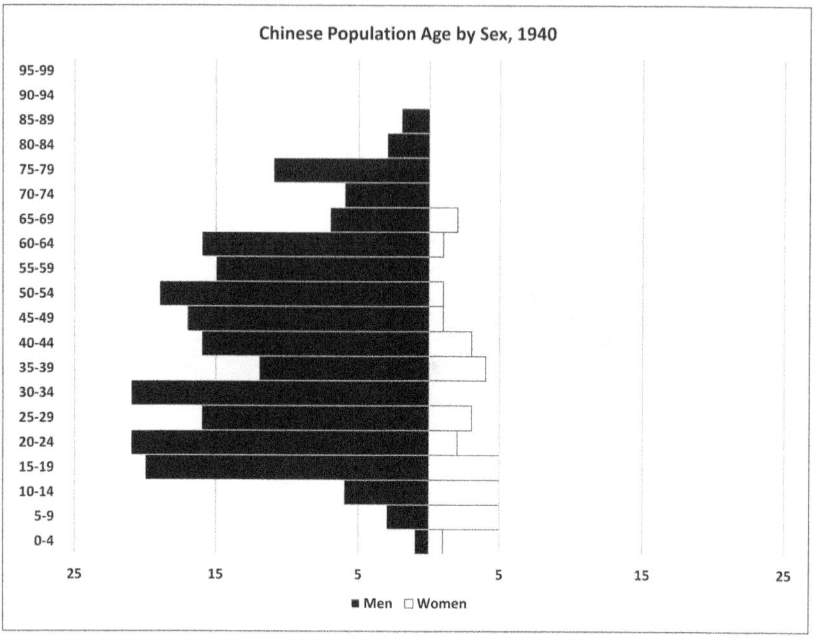


Figure 20: Population pyramid of Chinese in Montana, 1940. (Created by author.)

Table 20. Montana Chinese population by location, 1900 Federal Census

| Location | # | Location | # | Location | # |
|---|---|---|---|---|---|
| Butte | 277 | Red Mountain | 6 | Fort Shaw Indian School | 1 |
| Helena | 189 | Blackfeet Reservation | 6 | Harlem | 1 |
| Hellgate | 109 | Horse Prairie | 5 | Forsyth | 1 |
| Billings | 84 | Cottonwood | 5 | Philbrook | 1 |
| South Butte | 61 | Township 3 | 5 | Lavina | 1 |
| 10N3W (North Helena) | 55 | Bremer | 5 | Monforton | 1 |
| Bozeman | 49 | Gardiner | 5 | South Bozeman | 1 |
| Belmont | 49 | Dupuyer | 5 | Red Lion | 1 |
| Union | 47 | Columbus | 5 | Princeton | 1 |
| Smead | 46 | Big Sandy | 4 | Fort Harrison | 1 |
| Dillon | 44 | Highwood | 4 | 15N6W (Roger's Pass Area) | 1 |
| Kalispell | 43 | Fort Keogh | 4 | Madison | 1 |
| South Philipsburg | 42 | Garnet | 4 | Norris | 1 |
| Fort Benton | 39 | Township 4 | 4 | Rochester | 1 |

| | |
|---|---|
| 10N4W (West Helena) | 32 |
| Boulder | 31 |
| Thompson Falls | 29 |
| Livingston | 26 |
| Havre | 25 |
| German Gulch | 25 |
| Hamilton | 24 |
| Anaconda | 17 |
| Big Timber | 17 |
| Ophir | 15 |
| Chinook | 14 |
| Warm Springs Insane Asylum | 13 |
| Deer Lodge | 12 |
| Gold Creek | 12 |
| Lewistown | 12 |
| Miles City | 11 |
| Pony | 11 |
| Walkerville | 11 |
| Jocko | 10 |
| Sunrise | 9 |
| Township 7 | 9 |
| Cedar Creek | 9 |
| Drummond | 8 |
| Township 5 | 8 |
| Jefferson | 8 |
| Bannack | 7 |
| Silver Bow | 7 |
| Hecla | 6 |
| Townsend | 6 |
| Laurin | 6 |
| Musselshell | 4 |
| Frenchtown | 4 |
| Hunter's Hot Springs | 4 |
| Choteau | 4 |
| Lima | 3 |
| Radersburg | 3 |
| Lombard | 3 |
| Troy | 3 |
| Manhattan | 3 |
| French Bar (10N1W) | 3 |
| Flathead Reservation | 3 |
| Ward | 3 |
| Barrets | 2 |
| Black Tail | 2 |
| Glendive | 2 |
| Columbia Falls | 2 |
| Granite | 2 |
| BiMetallic | 2 |
| Trail Creek | 2 |
| Chico | 2 |
| Victor | 2 |
| Argenta | 1 |
| Red Rock | 1 |
| Glendale | 1 |
| Diamond City | 1 |
| Toston | 1 |
| Clark's Fork | 1 |
| Red Lodge | 1 |
| Belt | 1 |
| Sun River | 1 |
| South Boulder | 1 |
| School District 29 | 1 |
| Township 14 | 1 |
| Muir | 1 |
| Horr | 1 |
| Fridley | 1 |
| Corvallis | 1 |
| Stillwater | 1 |
| Marias | 1 |
| Saint Mary's | 1 |
| North Billings | 1 |

Table 21. Montana Chinese population by location, 1930 Federal Census

| Location | # | Location | # |
|---|---|---|---|
| Butte | 154 | Poplar | 3 |
| Helena | 66 | Big Timber | 3 |
| Billings | 48 | Deer Lodge City | 2 |
| Miles City | 18 | Cascade | 2 |
| Havre | 16 | Philipsburg | 2 |
| Lewistown | 14 | Bozeman | 2 |
| Montana State Penitentiary | 11 | Hamilton | 2 |
| Warm Springs Hospital | 10 | White Pine Vicinity | 2 |
| Blackfeet | 10 | Noxon Vicinity | 2 |
| Livingston | 10 | Ninemile Vicinity | 2 |
| Missoula | 10 | Glasgow | 2 |
| Anaconda | 7 | Columbia Falls | 1 |
| Kessler | 7 | Whitehall | 1 |
| Chinook | 5 | Trinity | 1 |
| Troy | 5 | Barret | 1 |
| Prickly Pear | 5 | Jefferson | 1 |
| Cottonwood | 5 | Superior | 1 |
| Floral Park | 5 | Hunter's Hot Springs | 1 |
| Scobey | 5 | Ophir | 1 |
| Lewis & Clark County Poor Farm | 4 | Trout Creek Vicinity | 1 |
| Baker | 4 | Shelby | 1 |
| Dillon | 3 | Liberty | 1 |
| Lima | 3 | Wolf Creek | 1 |
| Glendive | 3 | Choteau | 1 |
| Rexford | 3 | Cut Bank | 1 |
| White Sulphur Springs | 3 | | |

Table 22. Montana Chinese population by location, 1940 Federal Census

| Location | # | Location | # |
|---|---|---|---|
| Butte | 85 | Kalispell | 2 |
| Billings | 48 | Chinook | 2 |
| Helena | 41 | Baker | 2 |
| Livingston | 8 | Longview | 2 |
| Havre | 7 | Sidney | 2 |
| Miles City | 7 | Manhattan | 1 |
| Glendive | 5 | Laurel | 1 |
| Floral Park | 4 | Orchard Homes | 1 |
| Cold Springs | 4 | Twin Bridges | 1 |
| Great Northern Hotel | 4 | Browning | 1 |
| Missoula | 3 | Cascade | 1 |
| Prickley Pear Valley [sic] | 3 | Columbia Gardens | 1 |
| Bozeman | 3 | Deer Lodge | 1 |
| Anaconda | 3 | Glasgow | 1 |

as this family intentionally attempted to shed the vestiges of Chinese culture, including the visiting of family graves at Forestvale Cemetery.

The remnant Chinese population of Montana demonstrated a patriotic spirit during World War I. During the initial volunteer draft registration for World War I, which was an unpopular action taken by President Woodrow Wilson, 117 Montana Chinese men provided their information. According to the draft regulations, aliens to the United States were required to register but were not included in any of the selection pools. The 117 registrants reflect a substantial portion of the adult male Chinese population in Montana, as the 1920 census listed 872 total. Butte provided the highest number of Chinese draft registrants with 39. In total, Chinese in 26 Montana counties registered for the World War I draft.

In Sweet Grass County the local Chinese organized a War Savings Organization, which sought to raise money for the war effort. Tom Kue, a prominent Big Timber Chinese mentioned earlier in this chapter, was the president of the organization. Even though Kue lacked American citizenship, he organized 15 other Big Timber Chinese and promoted the organization, raising a substantial amount of money for the war effort. The national office for the War Savings drive used a photograph depicting the members of the Sweet Grass County society as a promotional tool and sent the image around the United States (*Mineral Independent* 1919:4).

### *Departure and Death*

Stories on the death or emigration of aged Chinese individuals dominated the coverage of Chinese people in Montana's newspapers in the first four decades of the 20th century. Before 1900 most stories on the Chinese had focused on the threats of their immigration or continued occupation of Montana and the negative aspects of their community. In the 1900s the stories changed to public interest pieces on the death of aged Chinese individuals or the parties thrown for their departure to their homeland in China. The return to China before or after their death was a powerful motivation for Overseas Chinese, as beliefs held that without friends or relatives to worship at one's grave, the spirit of the deceased would forever linger in a type of limbo (Rouse 2005:22). Emphasis on the spiritual need to return home was not the only motivation, as many Chinese in Montana had not seen their parents, wives, husbands, or children for decades (Chung and Wegars 2005:2). Among Chinese women in Montana in 1900, 64% were married, all to Chinese men within Montana.

That same year 41% of Chinese men were enumerated as married (698 of 1,684), two were divorced, and another nine were widowed. By 1930 there were 17 married Chinese couples in Montana, with 44% of Chinese men married (183 of 414). In 1930 all but one of the married Chinese women were residing with their spouse, while 90.7% of married men were not living with their wives, who were likely still in China. Most of these Chinese men had left their wives and children in China, creating a sizable proportion of the population without their spouses.

Chinese in Montana retained contact with friends and relatives in China through letter correspondence. Examples of Chinese correspondence with relatives are rare in Montana, though a letter found in the Montana Historical Society Archives from 1893 provides some context to the types of topics covered. The letter to Ah Hei in Montana came from his older brother, Hao Hsing, in China, and attempted to bolster the spirits of the overseas immigrant. After hearing that the Ah Hei was ill, Hao says, "If you make even a little money, come home right away in order to restore your health. Do not overtax yourself by being away, so that mother and the family elders will not worry about you." Later, the elder Hsing relates a traditional Chinese story about the dangers of focusing too much on fame and wealth. Hao Hsing ends the letter with some advice grounded in both spirituality and familial bonds:

> *You should not be preoccupied with worldly things. Do not become too attached to foreign places, you should consider returning home to China your most important goal. I trust the gods will protect you and your good health. When you get enough money to come home, find a fellow countryman who will help you return quickly so that mother and the family elders will not have to worry about you, so that they are unable to eat or sleep. (Hao Hsing 1893)*

The wish to return home was a constant goal for many Chinese in the United States and other overseas communities, especially as their U.S. population numbers shrank. This social imperative helped coalesce the image among the American public of the Chinese as sojourners (Chung and Wegars 2005:2). Given the forces in China that drove them to immigrate (discussed more fully in chapter 5), it is unclear whether this social imperative would have been as strong had the immigrating Chinese been able to bring their families. The death of Mar Joe's wife and two children in a 1938 Japanese bombing raid on Guangdong while he lived in Thompson Falls is a poignant reminder of the heartache caused by America's restrictive legislation (*Helena Independent* 1938a:3).

Choteau residents celebrated the 1933 departure of Soo Son, who lived in Teton County for 44 years. Soo Son came to Montana in the 1860s, working mining claims in Helena before being hired by the Sands and Taylor Cattle Company, and then finally arriving in Choteau in 1889 and establishing a restaurant (*Helena Independent* 1933b:3). It is likely that Soo Son knew George W. Taylor, mentioned earlier, who was adopted by Jesse Taylor of the ST (*Mineral Independent* 1930:4). Numerous dignitaries came to Son's farewell party, including one of his old friends, Senator John E. Erickson of Choteau (*Helena Independent* 1933b:3). Soo Son had retired from his restaurant a few years before his farewell party due to increasing age and poor health, and his decision to return to China was centered on the wish to join his family. Similar to Tommy Haw, Soo Son never received American naturalization, though all dignitaries proclaimed him as an example of a model citizen (*Helena Independent* 1933c:3). At the conclusion of the dinner and speeches, the guests presented Soo Son with several gifts, including money for travel, and "tears welled into his eyes as his friends gathered about him to pay him tribute" (*Helena Independent* 1933c:3).

Other elderly Chinese left Montana during the early 1900s, including the former prominent citizen of Big Timber, Tom Kue. In 1924 Tom Kue left Big Timber and first stopped at San Francisco before heading by steamship back to China. It appears that Kue's purchase of a larger restaurant in the 1920s bankrupted him, and forced him to sell his holdings (*Helena Independent* 1924:3). The pattern continued across Montana, as Sam Wong, 71 years old, left Forsyth in 1925, Wong Ching left Helena in 1926, and four elderly Chinese men left Butte in 1927 (*Helena Independent* 1925b:3, 1926b:3, 1927a:3, 1927b:3). Others left Montana for other places in the western United States (*Montana Daily Record* 1908:3). Stories following their departures painted them as model citizens who were heading home or elsewhere in the West, only to die. The four Chinese who left Butte in 1927, aged 70–84, had apparently worked the placer claims in Alder Gulch in the 1860s. Remembering a different version of history than actually occurred, the newspaper story notes that the Chinese men "came in on the tail end of the placer mining excitement of the 60's when the presence of a Mongolian was not resented by the red-shirted miners" (*Helena Independent* 1927b:3). It was common in these editorials to paint the life of the Chinese as extremely trying and difficult, with the only adversity in their lives coming from their own inability to work well-paying mineral deposits.

Throughout the early years of the 1900s, it appears that even more of the original Chinese pioneers of Montana died than emigrated. In Helena, stories followed the deaths of local Chinese men including Toy Ket and Yee Chuck. Unable to work due to poor health, and subsisting off the food found in restaurant and hotel trash barrels, the 66-year-old Toy Ket died of a stroke while awaiting a transfer from his home to the Lewis and Clark County Poor Farm (*Helena Independent* 1928l:3). Yee Chuck, also known as "Ice Cream," died in Helena after serving the frozen dessert out of a two-wheeled cart for over 45 years (*Helena Independent* 1928n:3).

Smaller communities like Dillon, Philipsburg, Blackfoot City, and Deer Lodge documented the deaths of the last Chinese residents in their communities. Ah Sue, who was reported to be nearly 100 years old, was found dead in his small mining cabin in the Blackfoot City area. Sue had come to Montana in the mid-1860s, prospecting along Ophir Creek, and remained long after most others abandoned the area (*Helena Independent* 1929c:3). Tom Yee died at Philipsburg in 1932 and was the last remaining Chinese individual in that community. Yee was a member of the Methodist Episcopal Church in Philipsburg and was buried in the city cemetery (*Helena Independent* 1932:3). In 1934 Quong Lun Kee died in Dillon and spent a year in storage awaiting sufficient funds for shipment to China for burial (*Helena Independent* 1935:3). The Dillon Chinese community reportedly ended in 1938, with the death of Jim Lee. Known locally as "Whiskey Jim," Lee came to Dillon in the early 1900s to oversee the shipment of Chinese bodies back to China and decided to stay in the community. It is unclear whether Jim Lee's body ever returned to China (*Helena Independent* 1938b:3).

The last Chinese resident of Mineral County died in 1922 (*Mineral Independent* 1922:4). Local oral histories provide a diverse array of possible ways in which Cedar Creek John, as he was commonly known, arrived in the Superior area. All stories note that John arrived in the Cedar Creek Mining District during its peak, but he was variously identified as a doctor, a packer, or even a murderer repaying a debt to the victim's family in China. After the boom of Cedar Creek passed, John remained in Mineral County and sometimes watched the local children. According to oral histories, John watched children in the late 1890s and early 1900s for local residents and taught them how to use chopsticks. After spending a short stay in Missoula's St. Patrick's Hospital for an undisclosed illness, Cedar Creek

John returned to Superior, where he died at age 90 at the Ordean Hotel (*Mineral Independent* 1922:4). Demonstrating the still prevailing opinion of the Chinese, the *Mineral Independent* obituary led with, "Cedar Creek John, the only welcome Chinaman in Mineral County, died Monday morning" (*Mineral Independent* 1922:4).

### *Chinese Cemeteries*

The predominant narrative regarding the treatment of the Chinese dead is the shipment back to the ancestral homeland. While in his 80s, Henry Reslip recounted a story in the 1930s about an encounter with a Chinese person and a willow basket during the late 1880s. When Reslip asked what the Chinese man had in the willow basket, he replied, "Two Chinamen, one China lady" (Works Progress Administration [1930s], item 72:2). Reslip continued: "The bones of the three had been buried a little above the . . . site of Louiseville" (WPA 72:3). In another instance the bodies of 28 Chinese who died in Chicago between 1879 and 1887 passed through Helena on the Northern Pacific Railroad in 1894, each sealed in an airtight casket (*Helena Independent* 1894a:3). Many of the Chinese who died in Montana remain in their graves, though the predominance of removal narratives among the public and some historians continues.

Far from home, the death of a member of the Chinese community took on extra significance and became an occasion for displaying social solidarity. As illustrated by Kuijt (2001:82), "The power of ritual as a cohesive force is based, in part, on the realization that mortuary practice is a form of public action . . . often to elicit community participation." In this framework mortuary rituals are also a form of community maintenance, and others (Durkheim [1912] 1995; Gluckman 1954; Marshall 2002) support this view. The place of burial for the Chinese dying in Montana varied, with NPRR workers generally buried along the grade (Fullerton 1961), while others used established cemeteries. As in other places in the American West (see Chung and Wegars 2005), the Chinese interments at official cemeteries in Montana occurred along the fringes of the grounds. There are confirmed Chinese burial plots in several Montana communities including Missoula, Helena, Butte, Philipsburg, and Deer Lodge. Missoula and Helena's cemeteries shifted over the years, with many of the original plots now covered in urban development. For instance, the original Missoula city and county cemetery was located along Lower Rattlesnake Creek but was abandoned by the 1890s for residential develop-

Figure 21: Brick and mortar Chinese funerary burner at Forestvale Cemetery. (Photo by author, 2007.)

ment. As late as 1900 the Chinese community was still using the area for burials, including Tom Gogk, who died at the Missoula Joss House, and "Old" Charlie, who worked at the gardens (*Anaconda Standard* 1900a:12, 1900b:12). Today the original Missoula cemetery lies underneath dozens of tidy bungalow homes, and occasionally human remains appear during excavation for garage foundations and tilling of the ground for gardens. Some of the human remains appearing in the last few decades include the Chinese, underscoring that not all bodies were returned to China.

Some of the most significant aspects of traditional Chinese funerary rites included the offering of foods and drinks and the burning of paper money and personal effects to aid in the rest of the decedent's spirit. To maintain this cultural tradition, Chinese mourners constructed large brick-and-mortar funerary burners in the center of cemetery plots in Helena, Butte, and Philipsburg (Figure 21). While Philipsburg's burner is in complete disarray, the Butte and Helena examples remain in remarkably good condition. There are likely other examples of funerary burners in Montana, though in poor condition due to the ravages of time and the

Figure 22: Marble headstone for Chinese grave at Forestvale Cemetery. (Photo by author, 2009.)

impact of vandals. The Granite County Historical Museum in Philipsburg is looking into restoring the Chinese funerary burner at the town cemetery and thus preserving this significant legacy for this cultural group.

Over the first 30 years of Helena's history, the city had two major cemeteries. In 1890 the first two Helena city cemeteries were consolidated into Forestvale Cemetery, approximately two miles north of downtown. The Chinese community of Helena began burying their deceased at Forestvale in the same year, though in an isolated corner of the property far away from the non-Chinese burials. Between 1890 and 1954, over 220 Chinese were interred in the far northwestern corner of the Forestvale property, with an unknown proportion exhumed for return to China (Bik 1993:13). In 1988 the Forestvale Cemetery received funding for rehabilitation and fencing its boundaries concurrent with completion of a National Register nomination. At this time, the cemetery board decided to exclude the Chinese section, known locally as "China Row," from the newly constructed fence boundaries, as they viewed it as an inactive portion of the grounds.

Today China Row at Forestvale is overgrown with weeds, and the marble headstones (Figure 22) and funerary burner are in a state of disrepair. While there is clearly evidence of disinterment by contracted Chinese bone collectors, it is unlikely that all human remains were removed, as believed by cemetery officials. For instance, the 13 bodies moved from the Lewis and Clark County graveyard to Forestvale have no names associated with them, and it is unlikely that relatives or bone collectors could have relocated those remains. Bik (1993:13) notes that only a small proportion of the Chinese graves at Forestvale show evidence of disinterment. Scattered around the location of the funerary burner are hundreds of Chinese and European American artifacts reflecting ritual offerings to the dead. During a revisit to Forestvale Cemetery in 2008 and 2009, the researcher located a variety of artifacts, including Chinese ceramics (celadon, "Four Flowers," brown-glazed stoneware), opium paraphernalia (opium pipe bowls, brass opium cans), and a variety of personal items, including buttons and pins. Archaeological investigations of other Chinese funerary sites in North America (Chung and Wegars 2005) illustrate that Chinese incorporated non-Chinese material into their ritual practices. During funeral rites, the Chinese employed these various artifacts for two main symbolic purposes: to ward off evil spirits or to provide nourishment or amenities for the deceased spirit (e.g., Rouse 2005:20–21). It appears that the Chinese mourners at Forestvale Cemetery placed European American artifacts with, and in place of, traditional material culture. Further research at Montana's Chinese cemeteries has the potential for refining issues of ethnic boundary maintenance, ritual, and social cohesion.

### *Yee Sing's Journey*

Among the many stories of death and migration during the early 1900s, perhaps the saddest story at the end of the period discussed in this chapter occurred in the mid-1920s. Federal authorities arrested Yee Sing, an 80-year-old Chinese man from Billings, for the sale of drugs. As this arrest was his third offense, the federal court in Great Falls sentenced Yee Sing to one year and one day of hard labor at the Leavenworth Federal Prison in Kansas (*Helena Independent* 1925a:3). On April 7, 1925, Yee Sing arrived at Fort Leavenworth Penitentiary with only $39.95 in personal money and no property. According to the Fort Leavenworth attending physician, Yee Sing suffered from bad teeth, poor vision, past effects of sexually transmitted diseases, and advanced senility. With all his health maladies, the

Fort Leavenworth physician recommended granting Yee Sing a first-floor cell. It also appears from Sing's health records that he spent most of his sentence ill, either in his cell or in the prison hospital (Fort Leavenworth Prisoner Files 1926).

Yee Sing served his entire sentence at Fort Leavenworth and was then sent by train to Great Falls, Montana. As Sing lacked the funds to pay for his own ticket, the federal government paid his fare (Fort Leavenworth Prisoner Files 1926). On March 3 Sing arrived at the Great Northern Railroad station in Great Falls (*Helena Independent* 1926a:3). Without any money, or a place to go, Sing sat motionless in the railway station for hours. Finally, when station attendants closed the facility they discovered Sing and called the sheriff. When the police arrived at the station they arrested Yee Sing for vagrancy. Upon arrival at the station they discovered his identity they provided him food and a cell for the evening. The last status of Yee Sing's journey was a telegram sent to a Chinese businessman in Billings (*Helena Independent* 1926a:3). It is unknown whether any friend or relative ever claimed the elderly and senile Yee Sing.

## End of Exclusion

After nearly 100 years of immigration restrictions, the repeal of the federal Chinese Exclusion Act and its descendants in 1943 signaled a change in U.S. and Chinese relations (Ling 1988:113). In a surprising and sadly ironic twist, the leading proponents of the repeal were politicians and labor leaders from western states including California, Oregon, and Washington (Li 2009:59). Those repealing the law stated that the reasons included bolstering Chinese morale, undoing a historic wrong, and creating goodwill for strategic and economic reasons (Li 2009:59). Even though the repeal had more to do with war strategy than social justice and immigration reform, the event changed the course of Montana's Chinese history. Chiang Kai-shek, leader of China in 1943, threatened to pull out of the partnership with the Allied forces in its opposition to Japan unless the United States government rescinded and repealed the unequal treaties and Exclusion Acts. The repeal was not complete and instituted a quota system (Li 2009:60), though it also allowed current Chinese residents of the United States to apply for naturalization. Repeal of the Exclusion Act came too late for Montana's once vibrant Chinese community, as the population reached a new low in 1950. Today there are as many Chinese in Montana as there were in the 1920s and 1930s; some families survived the tough period and live dispersed across the state. Where other immigrant groups flourished with open doors, the Chinese of Montana felt the full weight of legal and social infringements of personal rights and the community will likely never return to pre-1900 numbers.

# 4. Archaeology of the Overseas Chinese in Montana

Historical archaeology, the study of past peoples through the analysis of material culture and documents, plays a significant role in how scholars understand Montana's recent history.

Biases influence every aspect of historical documents, thus alone these data sources provide only a single perspective on an individual or event. Archaeology, which by definition comprises the waste material left behind by past peoples, provides a more personal and humanly unbiased data set. All archaeology focuses on human trash, waste, and abandoned goods, but these materials can result in important insights into the routine and daily life patterns of human behavior that are not well documented in the archival record. Cultural behavior that shapes this patterning of artifacts provides a personal glimpse into the lived experience of Montana's past inhabitants.

Not surprisingly, given the historical context of the Chinese in Montana, the number of identified Chinese archaeological sites decreased statewide in each subsequent historical period as defined in chapters 1 through 3 due to the decline in populations and the centralization of the remaining individuals in semi-urban Chinatowns. As most archaeology occurs on public lands, such as those managed by the Bureau of Land Management or United States Forest Service, the sample of archaeological sites available for research is biased but still important.

In the 1980s through 2000s archaeologists in Montana discovered several significant archaeological sites related to the Chinese experience. My research between 2006 and 2010 collated, highlighted, reanalyzed, and built upon this earlier wealth of information. Analyzing the archaeological material in concert with the construction of the first in-depth historical context for the Overseas Chinese experience in Montana allowed for a reflective discussion, whereby the historical narrative informed, and was informed by, the artifacts left behind by these early pioneer Montanans. By the close of 2010 University of Montana (UM) archaeologists had identified 40 archaeological sites with confirmed Chinese presence, disproven two supposed Chinese sites through archaeological excavations and survey, and identified an additional 20 sites with a high likelihood of Chinese archaeological deposits but no confirmation. Identifying these sites was time-consuming and likely not exhaustive but is a good start toward better understanding this population's contribution to state history.

Even though there were significant Chinese populations in the state of Montana from the territorial period through the mid-twentieth century, only five major archaeological projects focused on the Overseas Chinese experience: at German Gulch (Fredlund et al. 1991; Meyer 2001), Cave and Magpie Gulch (Rossillon 2002), Virginia City (Bowen 2004), Marysville (Hall et al. 2006), and downtown Butte (Rossillon 2008). This is a relatively minor amount of work compared to other states in the region, such as Idaho, Oregon, and Washington. Building off this foundational data, I conducted the excavation of five additional sites, recording of over 13 new sites, revisiting of another 50, and survey of several hundred acres of public and private lands, using field schools and volunteer-staffed projects. An additional thesis on excavated materials from German Gulch by William Norman (2012), plus other publications by Ellis and colleagues (2011), Merritt (2010), and Merritt and colleagues (2012) continue to build a better understanding of the Chinese experience through archaeology. Dedicated land managers, students, and volunteers have raised awareness for the archaeology of the Overseas Chinese experience in Montana, including new museum exhibits at the Mineral County and Granite County historical societies and the creation of two interpretive websites detailing the lives of these pioneers.

This chapter describes the various archaeological endeavors and results and further illustrates how archaeological data can yield a richer and far more personal window into the lives of the Chinese in Montana and beyond. The preceding three chapters included a synergy of archaeological and historical data, and more fine-grained detail is provided in the sections that follow.

### German Gulch, Montana (1864–1910s)

In the late 1980s the cultural resource management firm GCM Inc. conducted archaeological survey and testing of the German Gulch Mining District, located approximately 16 miles west of Butte, MT (Fredlund et al.

Figure 23: Chinese artifacts from German Gulch Archaeological Collection, owned by the Beaverhead-Deerlodge National Forest. (Photo by author, 2007.)

1991). The project resulted from a proposed road expansion to the Beal Mine at the head of German Gulch by Beal Mountain Mining Company and the Deerlodge National Forest. As discussed in chapter 1, German Gulch was a significant fixture in the early mining history of Montana, dating to the first discovery of gold in 1864 and the subsequent decades of placer and hard-rock mining. Chinese miners and entrepreneurs arrived in German Gulch shortly after gold discovery but dominated the area's population by 1870.

Archaeologists discovered more than 40 areas of mining-related occupations within the gulch, ranging from earthen dams to log cabins relating to both Chinese and European American occupations. Excavation of these features, including the remains of several intact Chinese habitations, recovered more than 30,000 artifacts relating to the life and habits of the German Gulch residents. Most artifacts were materials of Chinese origin (Figure 23). Subsequent road expansion destroyed the areas excavated by archaeologists, making the archaeological collection and associated documentation the only record of some of these areas. Much of German Gulch still retains significant integrity but reels from the continued destructive efforts of looters and vandals, and low-scale placer mining still continues along the creek's many intact mining claims.

Fredlund and colleagues' (1991) project result reports are excellent and detailed, though lacking in theoretical analysis with numerous instances of incorrect identifications of cultural material. Meyer (2001) used the results of these excavations in a forward-thinking master's thesis at the University of Montana, attempting to study the internal dynamics of the Chinese occupants of German Gulch. A major glaring oversight in both the project report and Meyer's thesis was a discussion of the final dispensation of the archaeological collection itself post-analysis. Discussions with Meyer determined that for his thesis he merely used the report data, not the actual artifacts, leading to further speculation on the whereabouts of this important archaeological collection. The Beaverhead-Deerlodge National Forest (BDNF) did not hold a copy of the finished report in its files. Finally, after nearly a year of effort, researchers discovered the collection in a wooden cabinet in the garages at the BDNF. The collection appeared in good condition, with the exception of poor curation over the last two decades. In 2009 the UM and the BDNF entered into a formal agreement for UM to house the collection temporarily to allow reprocessing and analysis while providing modern curatorial standards for bags and boxes.

With no intact or complete artifact inventory list, there was no means to determine the possibility of missing artifacts, though initial survey of the collection in 2009 identified that some of the most intact examples detailed in the report and featured in photographs were indeed missing. After months of dogged research, and by random pure luck, Dr. Priscilla Wegars from the University of Idaho and volunteer curator of the Asian American Comparative Collection provided a photograph showing most of the missing artifacts in a display at a gas station in Rocker, Montana, in 1999. After several uncomfortable phone calls to the gas station parent company with researchers trying to describe the cabinet full of opium drug paraphernalia (opium pipe bowls, cans, and other materials), archaeologists finally received notification that at least some of the materials were found in a box at the Mai Wah Society in Butte. Some artifacts are still missing from the collection, but the bulk of materials appear safe and in relative context. During the spring of 2010, 15 archaeologists and Passport in Time volunteers at UM spent one intensive week reprocessing the collection for further analysis. At the time of writing the reprocessed collection remains at the BDNF Supervisor's Office for future deposit at the Bureau of Land Management's Curation Facility in Billings, MT.

Unlike for many other Chinese mining camps of the early historical period, there is a wealth of historical documents regarding the German Gulch Chinese experience. These documents, however, do not include any first-person records from the Chinese themselves. Historical records instead come from deed books, mining claims, census records, and various local and regional newspapers. This being said, the German Gulch Chinese story is better understood than that of most other remote mining camps during this period, largely the result of archaeological interest in the site. While historians often focus on the written documentation as the sole source for discussion, archaeologists at German Gulch reconstructed the human environment through intensive survey and excavations. This work yielded a unique glimpse into the cultural landscape of German Gulch, as noted in Norman (2012:156):

> *Local population centers, such as localities 30 and 48 [Upper and Lower Town, respectively], were situated near resource extraction areas but are generally established in accessible, relatively flat areas where topography and location enable easier acquisition and distribution of goods. Outside of these local population centers, it appears that, based on artifact class distribution observed in this study, smaller satellite sites, such as work centers and residential dwellings, radiate outward into the surrounding landscape. It is probable that these smaller occupations were engaged in a reciprocal relationship with larger population centers in the gulch, enabling the movement of goods and raw materials.*

Archaeologists excavated only a fraction of the total archaeological potential for the sites in the mining district but provided some of the most interesting and unusual artifacts for the Overseas Chinese experience in Montana.

As mentioned earlier, site archaeologists identified 40 separate areas within the German Gulch drainage containing archaeological materials and termed each a "locality"; that is, a small area of human activity all related to a larger whole, in this case the German Gulch Mining District. Localities ranged in size from a single cabin to an entire community of Chinese dwelling remnants at Locality 48. Beal Mountain Mining Company's proposed road did not impact all of the identified localities, with most of the destruction occurring at Locality 48 and 30. It is not important to discuss all the archaeological localities identified in German Gulch, but some of these areas offer important glimpses into the life of Chinese immigrants in the late 19th century in Montana.

Archaeologists mapped and tested one area at Locality 24, which consists largely of ditches as well as a retention pond and dam likely associated with hydraulic mining and indicative of Chinese mining practices. Sometime after abandonment of the nearby mining operation, the dam breached, leaving an exposed cross-section. Excavators in 1988 cleaned the profile of the cross-section and determined that the dam was of tamped-earth design, similar to known Chinese techniques (Fredlund et al. 1991:59). Valentine (2002) recorded similar tamped-earth dams at American Canyon in Nevada in association with Chinese miners. This same tamped-earth construction is likely the way Chinese workers constructed the fireproof store in Deer Lodge discussed earlier (*New Northwest* 1872d:3). The excavation at Locality 24 provides sufficient evidence to analyze other historic dams across Montana for a specific signature of Chinese construction. Specifically, excavation of dams at Chinaman's Cove and Cave Gulch, discussed in the next section, might help to discover the origins of the mining features at those locales.

Locality 30 excavations recovered evidence of reoccupation of abandoned structures by Chinese miners, and possibly even unique terrace construction organized along traditional *feng shui* principles. In many of the features at Locality 30 there was a mixture of both Chinese and non-Chinese artifacts, though several features appear to have solely Chinese materials. Archaeologists recorded 21 features at this locality and tested 10 of these, with the majority of artifacts of Chinese origin. It is likely that Locality 30 is the Upper Town area noted in historical documents and that by the early 1870s had been completely taken over by Chinese miners as they worked the adjacent claims.

One feature in particular at Locality 30 appears to have served a specific and special function that researchers did not fully determine. The mysterious feature is composed of local rock made into a circle two meters in diameter, with a narrow entryway extending to the southeast and a small ground-level air vent on the side. Fredlund and co-workers (1991:92) interpreted the feature as either a charcoal kiln or a type of oven for melting gold, retorting amalgam, or even baking communal food. The feature's construction does not fit examples of known charcoal kilns or even Italian or Slavic ovens (Wegars 1991). There is another possibility, however, as evidenced by Chinese cremating their dead in German Gulch in 1874. According to a newspaper report, the first cremation in Montana Territory was conducted by

> *the Heathen element of German gulch, who one day last week cremated six of their deceased countrymen in the most successful manner. The process is this: The dead are disinterred, the flesh scraped from the bones and these are placed in a crucible and reduced to ashes, which, in this instance are to be packed in tin boxes, box of ashes separate, and sent to the Flowery Kingdom for a final interment. (New Northwest 1874g:3)*

Little is known of Chinese cremation techniques in rural environments, so the feature might have been some type of Chinese crematorium. Excavations of the rock feature recovered only a handful of cow bones from the surface of the units, with nothing but charcoal and wood debris in lower levels (Fredlund et al. 1991:93). While the lack of cultural remains within the feature could support a charcoaling function, the lack of artifacts could also indicate cremation, with complete removal of bones and associated artifacts.

Locality 48 is probably associated with Lower Town, the largest of the European American and Chinese settlements in German Gulch, with the main part of the original historical settlement located a few hundred yards downstream. Remining after the abandonment of the Lower Town area near the confluence of German, Norton, and Beefstraight creeks has effectively destroyed the downtown section, leaving only this large satellite part of town intact. Archaeological evidence from Locality 48 indicates that Chinese miners and entrepreneurs solely occupied this area and had a diverse variety of businesses located within its bounds. Road widening efforts in 1988 had the most direct effect on Locality 48, as the proposal was to destroy three to four features on the east side of the access road. As such, archaeological crews attempted to excavate fully three features in the direct path of development. The archaeological materials from two side-by-side features indicate that the excavators might have encountered a Chinese business section. The two depression features shared a common sill log that extended a length of nearly 15 meters, and excavators even uncovered intact floorboards (Fredlund et al. 1991:98–101).

Artifacts encountered at these two features and an adjacent trash dump further support the conclusion that this area served as a business district, as there were fragments of large, thick-walled globular jars, scales and scale weights for measuring goods and gold, and a wide variety of Chinese ceramics, opium cans, and other assorted imported materials (Fredlund et al. 1991:104). The presence at German Gulch of large, thick-walled Chinese stoneware vessels suggests a more substantial and important Chinese store and import business. Even extensive excavations at Butte's Chinatown and the Cedar Creek Mining District, both discussed later, recovered no thick-walled stoneware jars (Rossillon 2008). The faunal remains at the trash scatter adjacent to the two business features included domestic species commonly found at historical archaeological sites (pig, cow, chicken), but also wild game (snowshoe hare, grouse, deer), and imported species from China (sheephead fish, Chinese olives, flounder). This is the most diverse faunal assemblage found in any of the Chinese sites from Montana across all temporal periods.

On the north side of the road, the opposite side from the Chinese business area, archaeologists carried out less intensive excavations at several features to determine their function and association with the rest of the site. At one feature archaeologists discovered a trash scatter containing a dense accumulation of opium pipe paraphernalia, leading to an interpretation of the area as a pipe mender's home (Fredlund et al. 1991:104). Specifically, crews discovered an opium pipe saddle, opium pipe bowl fragments, five opium cans, and 11 opium pipe fittings, of which nine were inside one of the recovered opium cans. These artifacts coincided with triangular files and provide strong support for interpretation as a pipe mender's domicile.

While Fredlund and co-workers' (1991) report was excellent and thorough for its time, they did not attempt basic intra-site analyses. In particular, they organized the collection by material class instead of excavation unit or even feature, providing little means of intra-feature comparison. Garren Meyer (2001) made the German Gulch collection the subject of his master's thesis, though with mixed success. Meyer (2001) focused his research on an ambitious and ground-breaking attempt to discern ethnic differences within the Chinese community of German Gulch, but he found that there is far too little information in the literature to determine if there would even be any material culture differences within the Chinese community based on ethnicity. Ultimately Meyer's analysis lacked sufficient archaeological or historical data to support his hypothesis, but the attempt provides a beginning point for future analyses. It is also quite apparent that Meyer did not revisit or reanalyze the collection, save for the extensive shell casings assemblage, as evidenced by the poor condition of the materials observed in 2008.

Norman's (2012) thesis resulted from the reanalysis of the German Gulch collection excavated by GCM in the 1980s and early 1990s. This research provided more data on an intra-site analysis as well as an analysis that integrates class, gender, and ethnicity. A central point of Norman's reanalysis is a reorientation of traditional archaeological thought regarding Overseas Chinese sites, focusing on a China-centered core and a Montana periphery, turning many European-centered core-periphery models on their heads. Chinese pioneers in Montana were part of a larger system of immigration and commodity transfer that linked directly to the government, corporations, and social organization from China. In this analytical framework, Montana was a colony of the Chinese nation, supplying work for immigrants but an outlet valve for an overpopulated country and a source of new revenue to the core. While this perspective is not as developed as the European core-periphery colonial models, the Chinese in German Gulch experienced many of the same facets of capitalistic expansion as their European American counterparts.

As described in chapter 1, German Gulch represented a common pattern for Chinese settlements in the 1862–1880 period in Montana: a robust and growing Chinese population, spatially and culturally segregated from their European American counterparts, and focused on the extraction of mineral wealth through placer mining. Protection from outside influence and control by the presence of a large and cohesive community, complete with at least two stores and possibly a temple and social meeting hall, allowed the Chinese in German Gulch to retain a level of autonomy and separation from their European American neighbors. Even so, the Chinese relied on European American miners and mining companies for access to new placer claims, most of which were purchased at a higher than market rate. As was common throughout Montana in the 1870s, placer yields dwindled, and the mining industry shifted toward a model of heavily capitalized hard-rock mining; this excluded the Chinese from participating. While there were continued low-level Chinese placer mining operations into the 1880s within German Gulch, it appears that many of those pioneers moved elsewhere in search of work, perhaps to the railroad or in a different industry within the booming town of Butte.

### Cedar Creek Mining District (1871–1880s)

In August 1995 Timothy Light and Mary Horstman (now Mary Williams), working for the Lolo National Forest, surveyed and archaeologically tested portions of the Cedar Creek Mining District, near the Idaho border in western Montana. Light and Williams attempted to answer some basic questions regarding the identity of the known town of Cinkers as the historic town of Louiseville (called Louisville today in collective local memory) and to locate evidence supporting the presence of a Chinese population inhabiting the drainage during the last half of the 19th century. During the testing of Cinkers/Louiseville and China Gulch, these Forest Service archaeologists uncovered artifacts pertaining to the early historic period of Cedar Creek, including a significant collection of Chinese-related materials. Light and Williams's original analysis detected neither the number and extent of these Chinese materials nor the significance of these artifacts to the history of the Chinese in Montana and the United States.

From the historical record, Louiseville was the central settlement in Cedar Creek for the 1870 gold rush but quickly waned by the fall of that year due to better prospects farther up the drainage. At that time Chinese miners, fleeing restrictive ordinances in Idaho, arrived in Cedar Creek and apparently set up an encampment in a small east-west-trending side drainage quickly termed China Gulch. Forest Service archaeologists excavated several test units near a U-shaped hand-stacked rock hearth in China Gulch, and in a cabin and flat at the old remnants of Louiseville. Results of the 1995 excavations suggested that Louiseville possessed a substantial Chinese archaeological component, directly post-dating abandonment of the settlement by European American residents. Excavation units at the rock hearth in China Gulch provided no clear ethnic indicators to the 1995 excavators but did recover hundreds of small bone fragments and bottle glass, supporting the hypothesis that these features related to domestic occupations. After the initial 1995 analysis, the Louiseville and China Gulch collections remained at the Lolo National Forest Supervisor's Office in Missoula. In 2007 UM volunteers reanalyzed the collection and observed material traits missed by the earlier excavators and germane to the area's Chinese occupation. Reanalysis prompted the creation of a new report using the original collection and field notes (Merritt 2007) and spurred further archaeological investigations during the summer of 2007 and 2008 (Merritt 2009a). The 2008 fieldwork also led to the discovery of the hand-stacked rock hearths near the now disappeared town of Forest

Figure 24: Chinese "funs" trays from China Gulch in western Montana, owned by Lolo National Forest. (Photo by author, 2007.)

City at Mary Anne Gulch, 1½ miles farther up Cedar Creek from Louiseville and recorded as the Rowley Chinese Hearths.

The China Gulch excavations, while not recognized in 1995, recovered several modified opium can fragments termed "funs" trays, suggesting a strong affiliation with Chinese miners. In fact, archaeologists consider funs trays purely a Chinese material class, as they currently appear only in Overseas Chinese–related archaeological deposits (Felton et al. 1984:68–69; Ritchie 1986:387). It appears that the Chinese used these modified opium cans as standard weights for measuring out portions of opium for sale (Ritchie 1986:387), but the trays might have also found use as gold dust weights (Merritt et al. 2008:2). While the artifact is not as recognizable as the more common Chinese ceramics of celadon or brownware, funs trays provide sufficient indication of a Chinese presence (Figure 24). What makes the China Gulch assemblage of funs trays distinct is their quantity, with 11 in various levels of formation recovered between the 1995 and subsequent excavations in 2007. Before the China Gulch collection, archaeologists had located only a handful of funs trays from other excavations in the United States (Felton et al. 1984:68–69), with the bulk of known examples (116) found in New Zealand (Ritchie 1986:387; Wylie and Fike 1993; Merritt 2007:2). In fact, excavations at nearby Louiseville recovered no additional funs trays, suggesting that the accumulation at the hearth might indicate a distinct cultural occurrence. As noted by Ritchie (1986:387), the primary use of funs trays was in the sale of opium, and the predominance of these artifacts in a single hearth feature in China Gulch suggests that the location may have served as a hub for opium trade or smoking in the area; this activity could have supplemented the income of local miners and/or provided some of those individuals with temporary reprieve from the rigors of the mining lifestyle.

Two intact U-shaped, hand-stacked rock hearths are perhaps the most distinctive cultural features of China Gulch. These rock structures seem unique to itinerant or short-term Chinese occupations, as evidenced by other hearths located in California (Smith and Dixon 2005) and New Zealand (Ritchie 1986). Other examples of this style of hearth construction have been found on other Chinese sites in Montana, including German Gulch and Upper Cave Town. Archaeological excavation of a similar hearth found in 2008 at Yogo Town, located in the Little Belt Mountains of central Montana, found no associated Chinese artifacts. Further survey of Cedar Creek discovered six to seven additional U-shaped rock hearths near the former settlement of Forest City—these were recorded as the Rowley Chinese Hearths. It is unclear if the hearths are actually Chinese in construction due to the lack of archaeological excavation, though they are nearly identical to those in China Gulch. An archaeological investigation in Cedar Creek of these hearths has provided significant information on historical and cultural processes of Chinese miners living on a mining landscape in western Montana. It is likely that the hearths served as cooking areas with tents constructed close by to take advantage of radiative heating from the rocks during cool Montana nights. Comparative work by Ritchie (1986) in New Zealand did not discover any examples of these types of hearths in China itself.

The variety of Chinese ceramics located at Louiseville suggests a long-term occupation and the importation of a wide variety of goods directly from China. Archaeologists recovered fragments of several celadon-style ceramic serving bowls, two Bamboo-style bowls, and two Four Flowers vessels. Imported storage vessels, commonly termed Chinese brown-glazed stoneware, include at least four liquor jars, five wide-mouthed jars, two spouted jars, and a single ginger jar fragment. In addition crews uncovered several unglazed stoneware lids that would have sealed a wide-mouth jar (Merritt 2009a). As there was no evidence of the large, thick-walled Chinese globular jars commonly found in areas with a store, as in German Gulch, this might indicate that Louiseville supported a large Chinese population but local stores imported only smaller, more portable Chinese vessels that held less bulky items. Foods commonly contained in wide-mouthed or spouted Chinese jars included soy sauce, vegetable oil, dried fish, pickled duck eggs, and a variety of preserved goods. After consumption of the original contents, these vessels could have served any number of functions (Yang and Hellman 1996). While the vessels can only suggest indirect evidence of certain foods, faunal remains provide more direct evidence of menu items at Louiseville. Unlike at China Gulch, excavations at Louiseville uncovered several hundred large and identifiable faunal materials largely reflecting domestic pig, cow, sheep, and chicken.

The artifact assemblage suggests that Louiseville's occupation was concentrated on the south end of the site, as a Missoula-based hotel owner occupied most of the north part of town. A dense concentration of 15 opium cans located in a single excavation unit in the southern end of town might indicate the presence of a communal recreation area for the purpose of smoking opium. Excavators located only a few additional fragments of opium cans across the rest of the site, further supporting a special-use area. Several of the opium cans retained the original orange and black paper label with Chinese characters, though they are illegible. Interestingly, crews did not uncover any opium pipe bowl fragments in the same unit, possibly suggesting that the opium smoking occurred elsewhere on the site (Merritt 2009a:16).

Louiseville's diverse population, at least in comparison to other settlements in the immediate area, is apparent in several other artifacts discovered during the 2007 and 2008 excavations. Archaeologists discovered several fragments of a Chinese inkstone, used in traditional calligraphy. This is the first example of a Chinese writing stone located at a Montana archaeological site and indicates literacy among at least some of the Chinese population of Louiseville. Ritchie (1986:500–502) notes that inkstones found at Chinese sites in New Zealand probably correlate to educated individuals who served as letter writers for uneducated Chinese immigrants. In the same unit excavators located a cobalt blue glass syringe plunger (Merritt 2009a:19). Analyzed in concert, the inkstone and syringe plunger might be evidence of an educated Chinese physician plying his trade in Louiseville, though we may never know this for certain.

Perhaps the most archaeologically significant discovery during the Cedar Creek excavations was the potential starvation signature evidenced by the presence of small, calcined, and smoked faunal fragments from the hearth in China Gulch. In seven one-meter-square excavation units at a single hearth, crews discovered over 1,000 small (< 2 cm) fragments of faunal material. Most significant is the complete absence of any larger fragments of bone, and faunal analysis identified only one of these fragments to a species, a domestic pig, *Sus scrofa* (Merritt 2007:45). Meanwhile at Louiseville excavators recovered a wide variety of identifiable large cuts of beef, pork, and wild game. This disparity between China Gulch and other Montana Chinese sites like German Gulch (Fredlund et al. 1991) or Cave Gulch (Rossillon 2002) suggests different taphonomic processes resulting from the processing of faunal material during a starvation episode. Analysis of the bones from China Gulch by Dr. Shannon Novak and Dr. Meredith Ellis of Syracuse University confirmed a starvation signature to the remains. Statistical analysis of the faunal remains indicated a 99% similarity to the Donner Party's Alder Creek camp, a well-known starvation and hardship episode in Northern California. Further excavations in 2008 at another hearth in China Gulch retrieved a faunal assemblage almost identical to those of the 1995 and 2007 projects.

A full report on this discovery and its implications for other archaeological sites has been published in *Historical Archaeology* (Ellis et al. 2011). This is a poignant reflection on the hardships faced by the Chinese in Montana and other parts of the American West. It is hypothesized that the harsh restrictions of the Moose Creek Mining District ordinances passed in summer of 1870 forced Overseas Chinese immigrants to leave their homes and businesses with limited supplies and move to a new area in Cedar Creek. Without the infrastructure available in a more permanent community, the 100–200 Chinese arriving in Cedar Creek adapted to a situation with little cash or gold dust to purchase replacement supplies. Meanwhile, the Chinese continued to undertake the labor-intensive, physically demanding effort of placer mining during Montana's waning summer and autumn. Within only a few weeks of their arriving in Montana, the weather turned to winter conditions, further stressing the Chinese miners in Cedar Creek.

All the while, it appears that European American miners in nearby Louiseville dined on large cuts of beef and pork only a few hundred yards away.

Cedar Creek is not a black mark in the history of the Overseas Chinese in Montana; instead it represents successful adaptation to rapidly changing social and environmental conditions. While we may never know the true hardships faced by those Chinese miners in the fall and winter of 1870–1871, it is clear that by the following year this population dominated all facets of life and landscape in Cedar Creek. Erection of a Chinese store and temple during the 1870–1871 period signaled the arrival of a dynamic cultural group cleaning the gold dust out of the frigid waters of western Montana. Similar to what happened in German Gulch and most other rural mining districts, the 1880s witnessed a marked decrease in local fortunes, with the Chinese leaving the area for better markets elsewhere.

"Cedar Creek John" was probably the most famous Chinese citizen ever to inhabit Cedar Creek and the Superior area. According to local folklore recorded in the WPA oral histories, there is some variation in the details of Cedar Creek John's history (WPA [1930s], items 72:3, 75:6, 86:3). Most oral histories state that Cedar Creek John came to the area during the earliest days of the Cedar Creek gold strike, the early 1870s, and died in the early 1920s. Trefle Lacasse, a long-term miner in the area, stated that Cedar Creek John used to haul supplies over the Idaho-Montana border between Moose Creek, Idaho, and the Cedar Creek mines (Hahn 1986:33). Other oral histories state that Cedar Creek John was a doctor or that he came to America after killing another man in China and was paying his wages to the murdered man's family (WPA 72:3, 75:6). Wherever the truth may lie, it appears Cedar Creek John left an indelible imprint on the memory of many Superior residents. According to Hahn (1986:33), Cedar Creek John was a trusted member of the community and even refused to move to the Butte Chinese community after falling ill and being sent to Missoula for treatment. An obituary from the *Mineral Independent* (1922:4) reports his death:

*Cedar Creek John, the only welcome Chinaman in Mineral County, died Monday morning at the Ordean Hotel. He was an early settler of this county, spending years placer mining on Cedar Creek. According to papers found in his possession he was about 90 years of age. For many years he has been a county charge.*

Cedar Creek John now lies, presumably, in his grave at the Superior City cemetery. The Mineral County Historical Society recently erected a grave marker to indicate the location of his grave to commemorate one of the earliest pioneers of the area. Attitudes of the European American community toward the Chinese are summed up in the line earlier noted from Cedar Creek John's obituary in 1922, calling him "the only welcome Chinaman in Mineral County."

## Marysville Chinese Laundries (1870s–1880s)

In 2005 and 2006 Western Cultural Inc. (WCI) excavated several dozen excavation units at the site of two Chinese laundries in Marysville, Montana, located approximately 25 miles northwest of Helena. This project was tied to the widening of an existing gravel road for Montana Department of Transportation (MDOT) and allowed the first investigation of a Chinese laundry in Montana. Due to the scope of the mitigation work, excavations focused on the front of the two laundries, with no investigations of the backyard areas, which influenced the final interpretations. Mining magnate Thomas Cruse discovered gold and silver lodes at Marysville in the 1870s, though the town did not boom until the early 1880s (Hall et al. 2006:22). Sometime during this period Chinese entrepreneurs moved into the town, established two laundries on its eastern outskirts, and operated a garden at the north end that supplied fruit and vegetables to the miners (Hall et al. 2006:41–42). While the Chinese population of Marysville never reached appreciably significant numbers, they did provide badly needed services to the town's European American inhabitants.

Time constraints, coupled with limited funding, lead the conclusions of the Marysville excavation report (Hall et al. 2006) into a common pattern of discussing acculturation as evidenced by the domination of European American ceramics and food choices. The persistence of this theoretical orientation in historical archaeology, particularly on research of the Overseas Chinese, continues to limit the applicability and relevance of these studies (see Voss 2005). Archaeological work at the site of the two laundries uncovered thousands of artifacts relating to the Chinese occupation of this area. Due to the limitations of excavating only on the front parts of the laundry property, much of the potential domestic refuse from these occupations remains at the site and may affect conclusions about foodways. Hall and colleagues (2006) note that the most likely location for the majority of domestic refuse was behind the structures, in the unexcavated areas outside the scope of work. The crews located a handful of faunal remains that indicate pig and cow consumption, which is not an uncommon pattern for this period. There is no evidence at Marysville of the diverse assemblage of faunal and floral remains as discovered at German Gulch, which is likely the result of the isolated nature of this community, distant from a Chinese store.

Hall and co-workers (2006:156) note that there is some evidence that supporting James and colleagues' (1994:168) model for laundry types at Marysville. For example, artifact distributions indicated more leisure and ethnic artifacts clustering around one of the two laundries, suggesting that this laundry catered to a Chinese population while the other served the non-Chinese residents (Hall et al. 2006:156). The leisure activities noted (Hall et al. 2006:156–157) include gambling, reflected in the form of poker chips, traditional game playing with glass "Go" pieces, and the smoking of opium. None of these recreational activities, with the exception of "Go" or *wei chi*, is exclusive to the Overseas Chinese experience. Laundries within many of these isolated settlements served as a social and recreational nexus for the Overseas Chinese population, a place of quite repose and camaraderie outside the purview and scrutiny of the dominant European American population.

There were several unique discoveries at the Marysville's Chinese laundries that defy easy interpretation. Archaeologists uncovered three unusual artifacts: newspaper letter presses, a pewter pheasant, and a seemingly ritually disposed cat skeleton. First, crews discovered four wood-backed lead letter presses with figural images likely used in newspaper or flyer printing. According to Hall and colleagues (2006:150), the two most likely explanations for these artifacts were the printing of handbills or flyers for advertising the services of the Chinese laundries or similar media used to announce the visit of a Chinese dignitary that occurred in the early 1900s. The letter presses depict several different images, including one with a stereotypical scene of a Chinese man and rickshaw. Historical records do not indicate the presence of a newspaper office on the lots occupied by the Chinese laundries in Marysville, but the town did have a weekly newspaper for several years.

A second unusual artifact discovered at Marysville is a pewter pheasant figurine (Hall et al. 2006:169). Archaeologists also discovered a pewter pheasant figure at German Gulch in a domestic trash scatter, though of a slightly different style (Figure 25). Both pheasants were apparently attached at their base (or feet) to another object, possibly a pewter stein,

Figure 25: Pewter pheasant recovered from Marysville Laundry excavations. (Photo by author, 2009.)

decorative container, imported English bottle stopper, or something similar. Research into the significance of the pheasant in traditional Chinese culture provides little explanation for its association with these two sites, though Hall and co-workers (2006:169) suggest it might represent a symbol of office or authority. Eberhard (1988:233–234) notes that in traditional 19th-century China the pheasant could represent infidelity, insignia of an office, a board game, or even just a simple token of luck. A thorough review of other archaeological reports in Montana did not locate any additional examples from Chinese or non-Chinese deposits. This might merely represent a unique circumstance of two pewter pheasants found in relation to Chinese sites, or it might reflect a unique cultural pattern.

Finally, archaeologists discovered a nearly complete, articulated skeleton of a cat on the broken half of an earthenware plate. The skeletal remains of the cat show no butchering marks, suggesting it was not consumed but perhaps just simply buried or ritually offered outside the building (Hall et al. 2006:66–67). Analysis of artifacts and features associated with the cat skeleton remain inconclusive for purpose of the burial. Further research into other Chinese archaeological deposits may provide better clues to the reasons for the intentional burial, whether simply coincidental or ritualized.

Marysville's Chinese laundries provide significant information on the lives of these immigrants during the last decades of the 19th century. With placer mining declining as hard-rock ventures increased, Chinese residents of Montana shifted their economic attention to service-based industries such as laundries, restaurants, and gardens. In booming mining towns the European American bachelor male populations desired these services, and the Chinese exploited that need. By the end of the 19th century the Chinese established hundreds of laundries throughout dozens of Montana urban and rural communities. Located on the outskirts of downtown Marysville, the two Chinese laundries serviced an ethnically diverse mining community while providing stable employment to the immigrants until the boom faded. WCI and MDOT arranged for the donation of the Marysville Chinatown collection to the University of Montana's (UM) Department of Anthropology during the fall of 2006 after the final report was finished. The materials are now used as a comparative and teaching collection.

## Northern Pacific Railroad Camps (1881–1883)

Construction of the Northern Pacific Railroad (NPRR) through Montana in 1881–1883 is perhaps the single most salient achievement for the Overseas Chinese in the state. Historians estimate that more than 3,000 Chinese laborers worked on the NPRR from the Montana/Idaho border eastward to the capital city of Helena. Chinese laborers figured in the NPRR construction even down to the final epic hours when the NPRR pitted Chinese versus Anglo-Irish workers in a race to finish the last foot of rail at Gold Creek in 1883.

Even with the significance of the Chinese contribution to the NPRR, their story remains mute in the larger narratives of transcontinental railroad construction in Montana. Poor record keeping by the NPRR regarding Chinese employees leads to only a few historical records being available. Newspapers provide the bulk of this historical information, and these documents are reflective of the period in which they were written, dripping with anecdotal notations and racial stereotyping. Teasing out the facts from these records is trying, to say the least, but newspapers provide the bulk of historical information. Archaeology, on the other hand, provides a markedly personal and holistic picture of Overseas Chinese experiences in Montana.

Gary Weisz, a former Burlington-Northern Railroad employee, spent decades of his life dedicated to the investigation of the Overseas Chinese experience on the NPRR in Washington, Idaho, and Montana. With the majority of former Chinese occupation sites located on private lands still owned by the railroad companies, Weisz's (2003) detailed self-published research provides the only record of these areas in Montana and Idaho. Weisz contacted me during dissertation research, and the subsequent collaboration led to the formal recording of nine archaeological sites along the NPRR in Montana and a formal co-authored academic publication (Merritt et al. 2012).

Archaeological excavations in the 1990s at the town of Cabinet Landing, just west of the Montana border in Idaho, located the first evidence of the Chinese laborers on the NPRR. The *Weekly Missoulian* (1882b:1) notes that Cabinet Landing

> *is at present the leading town between Missoula and Spokane Falls, in point of business importance, and number of population. It is the centre [sic] of railroad construction at present, and the noise of blasting in the rocks, only a few hundred yards from the town, reminds the beholder of war-times, and the scenes of bombardment at Vicksburg. . . . The town has at present six stores kept by whites, and three by Chinamen; two China wash houses, not less than half a dozen eating houses, two lodging houses, partitioned off by sheets; and from sixteen to twenty saloons.*

Archaeological excavations at Cabin Landing by Landreth and colleagues (1985) found only a limited amount of Overseas Chinese cultural material but indicated that these laborers lived in a spatially segregated part of the town, apart from the European American laborers.

Four of the archaeological sites identified by Weisz were parts of three named towns that have long since disappeared: Thompson River, Weeksville, and Eddy. A fifth large encampment was termed "Last Chance" by the NPRR workers and officials due to its location at the edge of the alcohol-free zone of the Flathead Indian Reservation to the east. As the NPRR moved eastward into Montana from Idaho, the leading end of the rails served as temporary encampments for both Chinese and non-Chinese workers alike. Several hundred to perhaps upward of 1,000 workers clustered at these end of line camps, packed into makeshift shelters fabricated from boxcars or in long lines of yellowed canvas tents paralleling the grade. The NPRR supplied most goods for these towns, but private concessionaries fulfilled workers' needs for alcohol, prostitutes, and other non-standard supplies.

From evidence in the archaeological record it appears that the Chinese also established an articulated and elaborate supply system to provide workers with a diversity of goods imported directly from China. Imported Chinese artifacts from these sites include Bamboo, Four Flowers, and celadon ceramic vessels, brown-glazed stoneware storage vessels, opium pipe bowls and lanterns, and railroad-specific tools. Wrobleski (1996:47) also noted the presence of Bamboo and celadon-style ceramic vessels at Chinese railroad camps along the Virginia City and Truckee Railroad in Nevada, constructed in 1869. At both Montana (Weisz 2003) and Nevada (Wrobleski 1996) camps, Bamboo wares dominated the ceramic tableware assemblages, as that style was cheaper than other available types (Sando and Felton 1993:165) and is an indicator of the lower economic status of these railroad workers.

Weisz (2003) also discovered a variety of Chinese and Vietnamese coins at both these sites, though the Chinese examples were better preserved due to their higher copper content. It is significant to note that Chinese and Vietnamese coins share highly similar phenotypic characteristics, and likely served the same function in Overseas Chinese communities, in Montana and beyond. Chinese and likely Vietnamese and Japanese coins of similar size and shape served several functions in Overseas Chinese communities, including as gambling tokens, magical talismans, medicinal cures, and good luck charms (Akin 1992:64; see also Farris 1979; Olsen 1983).

Of the nine recorded archaeological sites, two are located beneath the impounded waters of a reservoir near the Idaho border, along an abandoned grade of the NPRR. These two sites yielded a broad variety of Chinese ceramics, including celadon and Bamboo pattern semi-porcelain vessels, as well as Chinese and Vietnamese coins and a miscellaneous array of opium paraphernalia and the general tools of a railroad laborer. Of the seven remaining Chinese railroad sites, only a camp east of Thompson

Falls, Montana, was revisited by the author in 2008. With a commanding view of the Clark Fork River, densely covered with waist-high grass, and protected by millions of blood-sucking mosquitoes thick enough to carry away small family pets, the Thompson River Camp contained tent platforms and dugouts of both Chinese and European American workers. This site also contains the only identified Chinese graves along the NPRR in Montana, though many shallow graves likely line both sides of the grade from Washington State to Missoula.

Little remains of the towns associated with four additional Chinese railroad sites in Montana, at East Eddy, Eddy, Weeksville, and Perma Bridge. Weeksville in particular was a prominent fixture in the 1880s western Montana narrative, known as "Bloody Weeksville" due to the number of murders and lynchings. Archaeologically, a large cast-iron wok recovered by University of Montana archaeologists in the 1970s or 1980s is the most significant artifact from these four sites; I came across it by chance in the basement of the Anthropology Department. A wok, apparently found in direct association with a small dugout, is the best illustration of the perpetuation of traditional cooking techniques among the Overseas Chinese railroad workers. Later railroad construction, widening of the nearby state highway, and other development has destroyed significant portions of these sites.

The final large railroad camp west of the Flathead Indian Reservation, Last Chance, established in late 1882 or early 1883, stretches for nearly a half mile along the banks of Clark Fork River and displays spatial segregation of Chinese and non-Chinese workers as evidenced by the pattering of cultural material. Unlike at other railroad camps along the NPRR, Last Chance archaeological investigations did not locate any evidence of opium paraphernalia and found only limited alcohol containers, suggesting that the NPRR perhaps exerted more control over its employees as they neared the reservation (Merritt et al. 2012:682). Unfortunately, given the location of this site in proximity to both a highly fluctuating river and two small communities, there has been a tremendous amount of damage to the cultural material through erosion and looting.

Two final sites, named "Fu Sang" and "Taklamakan" by Weisz (2003) during his reconnaissance work in the 1980s and 1990s, could possibly reflect the most intact and discrete Chinese railroad worker sites in western Montana. Jammed as these sites are between near-sheer cliffs, the railroad grade, and the several-hundred-foot-wide Flathead River, access to them is nearly impossible without aid of a boat. Weisz (2003) identified several dozen tent platforms, dugouts, and hearths at these two sites, in addition to some of the most pristine and diverse Chinese archaeological assemblages along the NPRR. Disturbance to these sites has been limited, and there is potential to locate intact archaeological deposits to assist in the reconstruction of the life of the Chinese railroad workers in Montana through formal excavation of these encampments.

From the Perma Bridge site on the border of the Flathead Indian Reservation to just northwest of Missoula, Montana, the NPRR used predominantly Mormon graders to allay the fears of the Flathead Indian agent about the potential effects of introducing alcohol-fueled European American roughs into a volatile political and social situation. Just east of Missoula, Chinese graders picked up the bulk of the construction effort for the remaining 40 miles to the area of the NPRR's Golden Spike; archaeologists have yet to document a single encampment in this stretch due to the lack of formal survey.

In collaboration with Gary Weisz, Dr. Kelly J. Dixon of the University of Montana and I completed the first historical archaeological peer-reviewed journal article on the Chinese railroad experience in Montana (Merritt et al. 2012). This article highlights how the Gilded Age in American history affected the railroads' choice of cheap labor, particularly Chinese labor, to complete the first transcontinental railroad through Montana. Later transcontinental railroads completed through Montana, namely the Great Northern Railway and the Chicago, Milwaukee, St. Paul and Pacific Railway, completed in 1893 and 1909 respectively, used large numbers of Japanese, Italian, and Slavic immigrants for the bulk of the construction labor force.

Overseas Chinese railroad workers contributed a significant amount of labor to the construction of the NPRR during the 1880s and continued as maintenance workers even into the 1890s and early 1900s. What is remarkable about this contribution is that the construction was occurring during the debate, and ultimately passage, of the federal Chinese Exclusion Act in 1882. This act led both active and newly constructed railroads to employ a different cadre of workers, including Japanese. Nevertheless, the NPRR construction in the 1880s led to a significant boom in the Overseas Chinese population statewide by the 1890 census, likely a direct result of these railroad workers entering Montana communities and the improvement of the state's economy.

Archaeological evidence helps to build a personal story of the trials, successes, and tribulations of the Chinese railroad workers in the late 19th century. The diversity and density of imported Chinese goods hints at the complex commercial network reaching from China across the Pacific Ocean to the rural and quickly moving railroad camps of western Montana. Hardships faced by the Chinese in Montana resulted in the death of dozens of workers, leading one newspaper reporter to comment: "Verily the road was built with Chinaman's bones" (*Helena Independent* 1884b:3). Both the NPRR grade and the archaeological sites along its length are testaments to the hard work and perseverance of the Overseas Chinese in Montana.

### Big Timber Chinese Laundry/Restaurant (1882–1948)

Archaeological evidence of the Overseas Chinese experience in Montana post-1900 largely exists within the boundaries of towns and semi-urban communities due to the demographic shift out of rural mining camps to centralized settlements. As the Overseas Chinese in Montana shifted their economic effort toward service-based industries such as restaurants and laundries by 1900, there was a centralization of population, as discussed in chapter 3. In order to investigate the archaeological legacy of Chinese during this period, researchers needed to focus on an urban or semi-urban setting. Justin Moschelle (2009), a master's student at UM, focused his research on his home community of Big Timber, located along the NPRR grade in south-central Montana. Under my direction, students and volunteers conducted a three-week archaeological investigation of the subsurface remains of a Chinese business that switched from laundry to restaurant during its operation. There was no surface evidence of the building; instead researchers used Sanborn Fire Insurance Maps to identify the potential location of the building within a commercial downtown block of Big Timber. Ground Penetrating Radar and focused test excavations helped to locate the potential buried remains of the building, and the formal field school opened up several dozen test units to expose what still remained of this early building.

Big Timber's Chinatown dates to the late 1880s, with the construction of the NPRR. Located one block south of the main NPRR line, the Chinese community never exceeded several dozen but included several businesses, gardens, and residences. Tom Kee, a prominent Chinese businessman in Big Timber, owned most of the Chinese block and was a fixture in the entire community. Most of Big Timber's Chinatown disappeared through urban redevelopment after the city's Chinese population left in the 1940s. In 1908 a fire starting from the NPRR line destroyed a large tract of Big Timber, which included this Chinese establishment (Moschelle 2009:36). Chinese owners apparently reconstructed the building in the same location; excavators located a series of intact and scorched floorboards at the bottom of a burn layer, evidence of the 1908 fire. UM excavations determined that only the northern and western foundation of the reconstructed laundry/restaurant remained intact after nearly six decades of post-abandonment urban development.

From the cataloging of the 35,000 artifacts recovered from the Chinese laundry/restaurant at Big Timber, it appears that most of the assemblage relates to the use of the building as a restaurant, though there are several laundry indicators. Materials included a wide variety of faunal remains: beef, swine, chicken, turkey, and even turtle shell. A variety of artifacts supplemented this faunal assemblage: bottled condiments such as ketchup, serving dishes largely of American and European manufacture, a host of alcoholic containers, and even scraps of painted molding from the building's interior. All help to construct a picture of the menu and ambience of this Chinese restaurant. Given the site's urban context, temporal control of the archaeological deposits is not absolute, but it appears that the restaurant served a predominant mix of Americanized Chinese food and even traditional American foods, such as prime rib and T-bone steaks. Serving dishes were largely non-Chinese in manufacture or appearance, but there were fragments of a few hand-painted Chinese soup spoons and celadon bowls. Finally, the painted molding indicates that the interior walls of the restaurant were a dark red with trim and molding apparently painted a bright golden hue. This is a traditional color scheme in many Chinese restaurants even today.

The building's use as a laundry around the turn of the 20th century was not as clearly recognized in the archaeological record. Laundry artifacts discovered at the site indicate the type of technology employed and the accoutrements of the business. Unlike in the excavations at Marysville discussed earlier, crews at Big Timber recovered two nearly complete bluing balls and a third severely deteriorated example from areas to the north and west of the laundry. Several glass fragments with a blue discoloration also indicate reuse of bottles to hold bluing balls for use in the laundry. Bluing balls (including powder and liquid forms) served both domestic and commercial laundry operations throughout the 19th century and into the mid-20th century, in order to make white fabrics appear whiter and thus cleaner (Sheppard 1909:50–51).

The only other laundry-related artifacts at the Big Timber excavation were several iron stove fragments, two of which relate specifically to a type of laundry stove that was used to heat flat irons. According to Sheppard (1909:72), such stoves work best in a situation where the laundry operator does not need the heat for another purpose, thus suggesting that the Chinese likely had at least two work stations within the building, one for ironing and the other for heating water for washing. These types of stove parts are linked directly to a commercial or subsistence laundry business and were relatively cheap compared to cooking stoves as shown in the Sears, Roebuck & Company catalog from 1897 (Sears, Roebuck & Co. 1897:138).

Between 1900 and 1920 a small wooden addition appeared on the northeast corner of the restaurant. Archaeological evidence, with some supporting historical data, suggests that this section of the building might relate to an opium-smoking area. Excavators located a cluster of opium-related artifacts including opium cans and pipe bowl fragments within this northeast portion of the building. It appears that this small outbuilding served as a possible opium-smoking establishment for the Big Timber Chinese (and perhaps non-Chinese) population. Opium-related artifacts discovered in this area of Big Timber include orange and burnished opium pipe bowls, opium cans, and possibly the remains of an opium lamp shade and wick support/base (see Wylie and Fike 1993:289). One opium can in particular had a Chinese character–embossed cartouche that translated as "Source of Beauty" and represents one of the most popular brands of opium during this period (Wylie and Fike 1993:287). Moschelle (2009:15) notes that in the 1930s local law enforcement officials raided a Chinese establishment in Big Timber's Chinatown, ransacked the building, and stole hundreds of dollars from the owner. It is likely that this raid focused on the building excavated by UM archaeologists.

Big Timber's excavations highlight the changing circumstances of the Overseas Chinese experience in Montana during the early to mid-20th century. These small urbanized Chinese populations within Montana's small communities represented the last vestiges of the state's dwindling Overseas Chinese residents. Mining had long since been eclipsed as a viable economic opportunity for the Chinese still living in Montana, and thus focus shifted toward service-based industries such as restaurants and laundries. Surrounded by a substantially larger European American population, Chinese restaurateurs in Big Timber apparently altered their traditional menu to meet the needs of the dominant non-Chinese population. With Tom Kee's departure from Big Timber in the 1940s, the Chinese history within that community ended, and it has only been resurrected through careful archaeological and historical investigations.

### Butte's Chinatown Excavations (1900–1940s)

Finally, the most prominent recent archaeological investigation of the Overseas Chinese experience in Montana stems from work conducted in Butte. Excavations at Butte's Chinatown (Rossillon 2008) cross-cut several aspects of Montana Chinese history and archaeology and represent the most comprehensive investigation of any site in Montana. In 2007 staff and volunteers excavated several features on a vacant lot adjacent to the Mai Wah Museum in Butte's historic downtown. The Butte excavations touched on significant themes in research, including household composition, occupation, socioeconomic class, and community maintenance and performance. This block of Butte's Chinatown illustrates the diversity of Chinese employment during the first part of the 20th century, with merchants, laborers, civic leaders, and all other walks of life represented. Archaeologists discovered several features relating to the household middens of Chinese families and the Chinese Baptist Mission.

Butte's Chinese Baptist Mission was founded in this location by 1900 and served as a central means of promoting integration of this population into the rest of the town's metropolitan masses. At first the Baptist Mission focused on providing education in English writing and language for the Chinese community's children (Rossillon 2008:15–16). As the population of Chinese declined in Butte, the mission's charge shifted toward acculturation, developing activities and organizations such as the first Chinese Boy Scout troop in the 1930s (Rossillon 2008:16).

The Butte Chinatown excavations recovered nearly 60,000 artifacts, though only 28,000 were retained due to redundant or insignificant information (Rossillon 2008:22). It is clear that the Butte Chinatown excavations recovered the largest variety of Chinese ceramics of any site in Montana, with 160 different vessels discovered (Rossillon 2008:83). Chinese ceramics include ginger jars, spouted jars, straight-sided jars, liquor jars, Four Flowers dishes and bowls, an Eight Immortals pattern bowl, and even examples of Japanese ceramics. The most significant absence among

Chinese ceramic types from the Butte Chinatown collection involves thick-walled stoneware vessels used for transport of dry or pickled goods (Rossillon 2008:86). As noted earlier, German Gulch contained several examples of these thick-walled vessels, and recent excavations at Coloma Ghost Town near Missoula recovered some additional examples. The presence or absence of these thick-walled vessels might be an important indicator of Chinese merchant activity, or perhaps a temporal marker, and warrants further investigation.

As noted by Rossillon (2008:71–72), there are differing levels of affluence associated with the trash middens in the Butte Chinatown block. The higher incidence of imported fruits, as evidenced by seed remains, and of higher-priced Asian ceramics does support a socioeconomic variation in households, especially for the largest trash midden excavated in 2007. Rossillon focuses on the presence of expensive and unique or imported artifacts as signals of socioeconomic status, though this perhaps masks some patterns within the assemblage. A more nuanced approach to the archaeological remains at Butte might uncover more appropriate gauges of affluence, as Van der Veen (2003) illustrates that quantity as well as quality of goods can demonstrate social and economic hierarchies.

Rossillon (2008:94) aptly compares the rural Chinese assemblages of sites excavated in German and Cave gulches and notes that there is a high degree of artifact diversity in the urban compared with rural contexts. This same pattern is also demonstrated in the study of Chinese assemblages in Arizona (Bockhorst 2003). Market availability deeply affected the types and varieties of materials available to the Chinese—or anyone—living in urban settings. Rossillon's (2008:94) conclusion that the presence of a higher variety of goods at Butte compared with German Gulch reflects higher affluence does not account for site-specific formation processes and places interpretive emphasis on equal market availability at both locales. In addition, there are some analytical problems, as the German and Cave Gulch assemblages date to the 1870s–1890s, while most of the recovered materials in Butte relate to the 1910s or later.

Perhaps one of the most significant findings, and unique for Montana, was the location of a variety of domestic artifacts that suggest the presence of Chinese children in the area. As reported in the 1900–1930 federal censuses, Butte had the highest number of Chinese families in Montana, including the presence of children under the age of 10. Excavations recovered over 80 porcelain doll fragments, marbles, and a small cup from a child's tea set. Excavations at Marysville tentatively suggested the presence of Chinese children (Hall et al. 2006), but the Butte excavations firmly recovered evidence of this sometimes hidden demographic. The work at Butte's Chinatown folds into a larger effort in historical archaeology toward the study of children and childhood (e.g., Baxter 2005), and information on Chinese youth seems highly significant to the discipline.

Due to the constraints to analysis and study found in a contracted mitigation project, the collections recovered during the Butte Chinatown excavations have research potential and could result in several theses or dissertations. Current understanding of the assemblage provides concrete data on urban Chinese populations in Montana, while also providing a glimpse at American organizations dedicated to acculturation (making the Chinese more European), as shown by the remains of the Chinese Baptist Mission. In addition, the Butte Chinatown excavations located the first firm evidence for the presence of Chinese children, and some information on the socioeconomic differentiation of Chinese households.

## Toward an Archaeology of the Overseas Chinese in Montana

Previous archaeological work in Montana yielded a diversity of site types and cultural material that formed a strong basis for the reconstruction and interpretation of Overseas Chinese experience in the state. Difficulties in finding the documentation, let alone determining the dispensation of the German Gulch collection, required years to overcome for this research. A comprehensive database of known archaeological sites, reports, and collections will aid future researchers in Montana and beyond to analyze the lives of the Overseas Chinese. Montana's archaeological record still possesses the potential for dozens of additional sites relating to the Chinese experience, in both rural and urban areas. Growing awareness of these types of cultural material, including educating archaeologists on identifying artifacts relating to this group, is reaping results by proper recognition of sites during survey and excavation. An interpretive website serves as a clearinghouse for information regarding the Chinese in Montana and is spurring better identification of sites and the interaction with all Montanans.

Archaeology provides a more personal glimpse into the lives of the Overseas Chinese in Montana and allows these past peoples some voice through the material objects they left behind. From the poignant starvation episode in China Gulch during the winter of 1870–1871 to the epic successes of the Chinese on the NPRR, archaeology is uncovering the forgotten history of this population. Damage to or destruction of early historic mining sites from remining of creeks during the early 20th century through industrial means such as dredging limits the potential for discovering intact Chinese communities from the 1860s–1880s. Further, urban renewal, forest fires, floods, and directed vandalism and looting reduce the number of intact Chinese archaeological sites each year in Montana. Identification of these sites is of paramount concern in the preservation or at least salvaging of important historic and archaeological information about this forgotten population.

Moving forward, scholars investigating the Overseas Chinese in Montana should focus their attention on significant and modern theoretical themes, such as adaptation, socioeconomics, intra-group interaction, trade and exchange, agency, and resistance, among other aspects. Old models focusing on the degree of assimilation and acculturation do not accurately reflect the realities of Overseas Chinese experience in Montana or the world. For instance, archaeologists identified an isolated placer mining operation dating to the 1890s to early 1900s, perhaps one of the last of its kind related to Chinese mining in the state. Of a diverse assemblage of artifacts found on-site, there was only one Chinese vessel discovered, but the historical documents demonstrated that the site was owned and operated by a company of Chinese miners. The paucity of Chinese ethnic markers found at the site does not indicate that this group of miners was more acculturated to European American practices but more that the closer consumer market for Chinese goods was nearly 60 miles distant. A dynamic view of Chinese life is far more appropriate for a group that was inherently dynamic and overcame all of Montana's social and environmental obstacles.

# 5. Chinese Social Organization in Montana and Archaeological Implications

## The Success of Overseas Chinese in Montana

While the traditional historical record does not view them as a successful immigrant population, the Chinese in Montana found a modicum of economic prosperity, community cohesion, and freedom. During the eight decades analyzed in this volume, the Chinese succeeded in pulling millions of dollars' worth of gold not only from the creeks and gulches of Montana but also from the pockets of European American miners and their families by providing needed goods and services. It is also apparent that this immigrant population banded together and vigorously fought the various legal challenges to their livelihood and citizenship in Montana. What allowed the Chinese to organize sufficiently in response to internal and external forces was the complex web of interrelations forged between the immigrant laboring population, the growing merchant-elite, family and clan relations in the homeland and in host communities, and a growing trend of nationalism.

Traditional South China social organization was centered around three separate modes: family lineage, socioeconomic class, and ethnicity (Woon 1984:1). By far, family lineage was the main means of organizing Chinese life prior to the 19th century. Upon reaching the far shores of new host communities, the traditional ties of family and clan grew strained over the long distance, and other means of social organization became prominent. This is not to say that ties to family and clan failed to translate into the diasporic community (see Speers 1870:546–547), but instead of being the main organizing force, these ties became part of a palimpsest. In 19th-century host communities Chinese established benevolent or mutual-aid organizations, clan associations, and secret societies with varying degrees of connection to the homeland. Members of clan associations could equally be joined to the district organization and possibly even hold a position in a prominent secret society all at the same time. Only in certain situations did these cross-ties become apparent.

The full depth of these interrelations is only now being integrated with mainstream historical narratives of the Chinese experience in diaspora communities; their impacts on archaeological and anthropological questions remain vague or untouched. Long-standing models for studying the Chinese experience focused on acculturation and assimilation, based on incomplete pictures of interrelations among members of this cultural group (e.g., Greenwood 1978, 1996; Blanford 1987; Longenecker and Stapp 1993; Diehl et al. 1998). It would be folly to investigate a community and its material culture without fully understanding the various components of the social structure, a point understood by anthropologists from functionalist, structuralist, or even Marxist approaches (e.g., Firth 1951; Radcliffe-Brown 1952; Levi-Strauss 1963; Bloch 1983; Godelier 1986). From an archaeological perspective, the interlaced web of social relations through new types of fictive kinship in diaspora communities deeply affected the site formation processes of Overseas Chinese sites.

Kerri's (1976:23) review essay on the historical and anthropological analysis of voluntary organizations illustrates that in changing social and economic situations immigrants employ these societies as adaptive mechanisms for creating new social groups in areas where kinship and territory-based organization fails. Specifically, Kerri (1976:34) notes that the study of voluntary organizations in anthropology shifted the disciplinary perspective so that researchers view "culture not only as an adaptive mechanism, but also as one of the things to which we as individuals and sometimes as groups must adapt." In new circumstances brought on by a variety of potential factors (industrialization, transnationalism, warfare, etc.), individuals adapted by employing these voluntary organizations as "innovative pathways to overcome the obstacles" encountered.

Little (1967:164–165) found that tribal migrants to South Africa from other inland regions with the inability to move upward socially used voluntary associations as a means of gaining power, prestige, and wealth. More recently, Brettell's (2005) study of Asian Indian immigrants in Dallas–Fort Worth discovered a similar pattern in the growth of voluntary organizations. Touching on a point raised by Kerri (1976:34), Brettell (2005:878–879) states that the immigrant population in the Dallas–Fort Worth area employed voluntary organizations at a higher rate as compared to other Asian Indian ethnic enclaves in Los Angeles. The main reason for this discrepancy, suggested by Brettell, is that the larger population in Los Angeles provided more opportunities for traditional kinship-based social organization than the under-sized and fragmented Asian Indian community in Dallas–Fort Worth. The Chinese use of voluntary organizations represents a similar adaptation to changing social circumstances, comparable to Little's (1967) analysis of secret society membership

Kerri (1976), Little (1967) and Brettell (2005) demonstrate the relevance and potential of voluntary organization research in anthropological and archaeological discourse. In particular, as Kerri (1976:33) notes, voluntary organization research demonstrates the failure of acculturation and assimilation studies to articulate true cultural systems fully due to the adaptive creation of new social organization in response to changing economic or cultural factors. As Voss (2005:432) and Voss and Allen (2008:19) discuss, the continued domination of assimilation narratives has marginalized Overseas Chinese archaeology, particularly in terms of the field's applicability to other disciplines. Archaeologists in Montana have continued this trend, applying acculturation models to their findings based on the presence or absence of traditional Chinese goods or their ratio to European American material culture (Fredlund et al. 1991; Rossillon 2002, 2003, 2008; Hall et al. 2006). Archaeologists studying the Overseas Chinese experience, as well as other ethnic immigrant groups, can use a framework of voluntary organizations to frame further research and refine ideas of how these associations deeply affected the daily lives of immigrants. The scholarly framework of the Overseas Chinese experience needs to include a more robust understanding of social organization in order to shed the well-worn and utterly bland vestiges of assimilation and acculturation studies. The first step is to discuss the varying webs of social relations within the Overseas Chinese community. This chapter presents an example of one type of social organization in Montana and then outlines an archaeological framework that can help guide future work in the field.

## Chinese Social Organization in Overseas Communities

As noted earlier, in traditional Chinese society the family and clan dominated community relationships. Due to the fractured state of 19th-century China, a distinct national sense of identity among Chinese citizens was lacking. Rather, in place of a national identity, the clan, class, and ethnicity formed the core of an individual's identity and role in Chinese society.

Upon reaching the shores of the diaspora countries such as Australia, New Zealand, Malaysia, England, Mexico, Cuba, or the United States, the Chinese, regardless of family, socioeconomic class, or ethnicity, became a monolithic entity in the view of the host community. The social pressure-cooker in host communities forced disparate Chinese groups into close quarters and cooperation for mutual survival and success. The historic perception of the Overseas Chinese as a monolithic cultural entity deeply affected the anthropological and archaeological analysis of this population throughout the 20th century. Recent years have seen a more robust and diverse variety of studies emerging (Ritchie 1986; Wegars 1993; Meyer 2001; Voss and Allen 2008), though much work remains.

Lyman's (1974) ground-breaking study of San Francisco's Chinatown establishes a better understanding of the "web of interrelations" that permeated urban and rural Chinese life in the United States and potentially other diaspora communities. According to Lyman (1974:473), group affiliation in Overseas Chinese communities fell into three basic categories: clans, benevolent groups (or district associations), and secret societies. As Zheng (2001:11) notes, each of the three types was based on a lineage arrangement commonplace in 19th-century China and thus comfortable for the immigrants to adopt. It is clear that membership in one did not preclude participation in the other two. Clan associations were the first institutions established by Chinese immigrants, with membership based on surname representing a break in traditional organization, which historically focused on blood relations. Clan associations in China possessed carefully constructed lineage charts to determine membership, though in overseas communities such evidence was hard to produce and ultimately became unimportant (Lyman 1974:477). Benevolent societies provided a larger net of membership, organizing Chinese immigrants by their district of origin, a common language, or ethnicity (Lyman 1974:479). Both clans and benevolent societies were extensions of legal entities acknowledged by the Qing government in China. A third organizing force in San Francisco and other diasporic communities was secret societies, an illegal institution in China due to their subversive tendencies toward the Qing government (Lyman 1974:484–485). Clans, benevolent associations, and secret societies helped to bind together an entire community faced by overwhelming social, legal, and physical attacks.

The clan, benevolent association, and secret society worked in unison and opposition in San Francisco's Chinatown during the 19th and early

20th century. Arguments can be made that these groups still deeply affect Chinese life in large urban centers in the United States (e.g., Parliament of the Commonwealth of Australia 1995; Bolz 1995; O'Neill 1999). In several instances the interests of the three groups conflicted, leading to what the host communities termed "Tong Wars" (Lyman 1974:490–491). There also appears to be a relative blurring in the distinctions among these three groups during the period in question, with merchant leaders of clans or benevolent societies also possessing interests in secret societies (Lyman 1974:491–492). Failure to accommodate the collective mandate of these three groups could lead to a member's ostracism, boycott, or exile not only from the organization but from a community (Lyman 1974:484). As seen during the Geary Act deliberations discussed in chapter 2, the Six Companies—a combination of several benevolent associations—deeply influenced the nationwide Chinese response to the law first through boycott and later through sponsorship of registrations.

Even with the supposed anti-Qing leanings of secret societies, the traditional social organizations in overseas communities focused on protecting their own membership and perpetuated a Chinese population that was fragmented in integration and purpose. By the late 1900s, the attempts to reform the Qing government in China helped to foster a new politically motivated and nationalist agenda among the Overseas Chinese that was nevertheless secondary to other loyalties. In 1899 the Dowager Empress Cixi halted any effort to reform the Qing government into a constitutional monarchy (Stanley 1996:475). Youwei Kang, the leader of the reformers, fled to British Columbia, Canada, and with the assistance of wealthy merchants established the Bao Huang Hui, or Chinese Empire Reform Association (Stanley 1996:475–476). In contrast to other voluntary organizations in overseas communities, the Chinese Empire Reform Association (CERA) cross-cut all lineage, class, or ethnic affiliations and bound most immigrants together in an effort to reform the homeland.

Liang Qichao, a student of Youwei Kang, visited the United States in 1903 when he was 30 to learn more about the republican values of the country, the better to reform the Chinese state. Due to his attempts at reform in China he was forced to flee to Japan in 1898 and traveled to Canada and the United States in 1903. Youwei traveled extensively in both countries, meeting politicians and local Chinese populations in order to learn about international forms of government and also to drum support for Chinese reform. During 1903 he visited Billings, Livingston, Miles

City, and Helena (Arkush and Lee 1993:81). Liang Qichao determined that China differed from America on a basic level, as Chinese were socially organized by clan rather than an individual unit. In effect, China had a village mentality, not a national mentality, something that America lost in the early 1900s (Arkush and Lee 1993:92; Stanley 1996:479). The creation of Chinese nationalism led to an active search for a new modern identity by those in China and overseas through active debate and sponsorship (Liu 2005; Chen 2006:192–193). Tsai (1986:92) notes that during Qichao's trip to the United States in 1903 the CERA expanded to include a total of 103 branches. Recent research by Dr. Weinong Gao (2009:443–445) indicates Liang Qichao observed that the Montana branches of CERA were the most organized and efficient and even created a separate division for mutual defense and self-protection.

Seemingly supporting the prevalence of CERA in Montana Chinese communities, excavators at Big Timber recovered several fragments of a decorative bowl made specifically for the 1904 World's Fair in St. Louis. The red transfer-print pattern shows the Hardscrabble homestead of President Ulysses S. Grant. Just before the 1904 World's Fair, C. F. Blanke purchased the Hardscrabble cabin for explicit use in marketing his new brand of coffee (National Park Service 2010). Blanke moved Hardscrabble to the World's Fair lands for the exposition and produced a variety of promotional goods that were sold or given away to visitors. These promotional items were not distributed outside the World's Fair, as indicated by historical sources (Bennitt 1905; National Park Service 2010). In 1904 Prince Pu Lun of China came to the United States to visit local CERA groups and visit the World's Fair in St. Louis. During this visit Prince Pu Lun met with leaders of CERA, and many joined his entourage to St. Louis (Bennitt 1905). While there are many other ways the Blanke bowl could have come to be deposited at Big Timber's Chinese restaurant, there remains a plausible connection to the St. Louis World's Fair and Prince Pu Lun's visit.

It is clear that all types of Chinese voluntary societies and organizations found favor with the Chinese living in Montana. There is no need to summarize the role and implications of all voluntary associations, as other scholars have touched sufficiently on the role and function of clans, benevolent associations, and CERA (e.g., Stanley 1996; Chen 2001; Jiang 2004; Bretell 2005; Liu 2005; Chan 2006; Chen 2006; Hsu 2006). As noted by Lyman (1974:474), many scholars avoided the role of secret societies in Overseas Chinese communities due to what they viewed as exaggerated claims of significance. The history of the Chinese in Montana includes numerous references to the role of secret societies in affecting daily life for these immigrants. To follow this thread, the following sections summarize the historical evolution of this facet of Chinese culture and its existence in Montana during the 19th and 20th centuries.

### Chinese Secret Societies in China

Secret societies, also termed voluntary associations, have a long tradition within Chinese culture, some dating back hundreds of years. In order to understand the evolution of Chinese secret societies, it is necessary to begin with the Han dynasty (206 B.C.–A.D. 220), as this represents the foundation for the cultural system that created millions of disenchanted peasants who formed the membership base of these organizations (Lee 1970:173). During the Han dynasty the ruling class began to institute formal examinations to allow individuals to enter into higher reaches of power. These examinations largely focused on the individuals' knowledge of Confucian texts and the deep understanding of their meanings (Smith 1994:60–61). Successful passing of the examinations allowed an individual to join the ranks of the gentry if the applicant passed only the first level of exams. Applicants became scholar-officials if passing second or third level exams. The ruling class of Chinese was more based on hereditary linkages than examinations. A small number of peasants acquired gentry positions through the state examination system (Grasso et al. 2004:14).

Confucianists adhered to the notion that humanity was able to be perfected through scholarly and academic study, and thus true power and position came from the ability to master the virtues of human life: loyalty, justice, benevolence, wisdom and propriety (Deng 1999:114–115). In this way passing the exams showed the dedication of an individual to the path of self-enlightenment that would be furthered through a lifetime of study. The gentry and scholar-officials were placed near the top of the traditional hierarchy via the examinations but below the ruling monarchy. An outgrowth of these beliefs was the low social hierarchy position that skilled physical laborers (miners, artisans, etc.) generally held within traditional Confucianist society.

Merchants bred distrust among the scholar-officials for one ideological and one pragmatic reason. First, in a system where intellectual pursuits were supposedly the single most important drive for individuals, the ever-present search for profit as exhibited by the merchant class was antithetical to this ideal. Second, the scholar-officials knew that the examinations, and thus the roads to power and prestige, were gated by expensive costs of training and education in doctrine, and the merchant class might well have the resources to enter the political system and begin to effect change (Davis 1971:12). The mix of ideological and pragmatic rationales in the Chinese Confucian system led to the mainstream exclusion of merchants, laborers, artisans, and many others, though some of these individuals did manage to move upward through the ranks of social stratification (Hsu 1967:300). While merchants were barred from gaining official local authority, the Qing used them effectively to manage tax collection (Mann 1987:12–13). Political and social opposition of the secret societies did not focus on Confucianism but on the Manchu and the Qing government.

While examinations were supposedly open to all Chinese citizens (except women), there existed a strong boundary between traditional power holders and the rest of the society (Smith 1994:82). These examinations were exclusive, based on access to schools, tutors, and examination centers, though millions did attempt and pass (Smith 1994:62–63). Not all villages or even provinces had dedicated centers for teaching and testing Confucianist ideology. In addition, the whole process of examination was cost-prohibitive to many, as every step required some outlay of funds to send a student to a tutor or travel across country to a testing center. This created an elite class that lasted the millennia due to the fact that a select few could afford to have their lineage become part of the Chinese system of power, while the vast majority outside that system rarely had a chance to enter (Ch'u 1962:161–163; Davis 1971:31). Those lineages that had attained these positions of power had to maintain their social and economic position continually through continued passage of exams by descendants, lest the lineage fall down the social ladder (Ping-ti 1962). For over 2,000 years this system perpetuated itself with only a few minor insurrections, due to the generally content agrarian population, though outside influences were beginning to take their toll. Economic growth and security and the relatively peaceful relations with neighboring states kept large-scale insurrection and rebellion to a minimum.

The Qing Dynasty began with the invasion and conquering of nearly all of modern China by the Manchu. The Manchu came from the north of China and, after conquering the nation, took over the system of Confucianism as a means of governance; some even suggest that the Manchu "were more Confucian than the Chinese" (Davis 1971:13). The new rulers attempted to decrease personal freedoms in order to quell any rebellions. These actions included portraying Buddhism and Taoism as dangerous (Stirling and Ward 1925:3; Smith 1994:83). Both those religions held the greatest sway over the densely populated agricultural hinterland of southern China and had direct influence in the development of secret societies (Stirling and Ward 1925:3–4). The Manchu edict forced these religious systems underground, increasing the friction between the new rulers of China and the bulk of the populace. Over the first few decades of the Qing Dynasty, Manchu families instituted themselves throughout the entire country as the new scholar-officials or gentry class (Spence 1990:46–47). In many cases non-Manchu lineages that had successfully passed the formal examinations were not included in the decision-making processes, leading to disenfranchisement (Brook 1990:28). As Manchu families controlled the examination system and the power-holding offices, they self-propagated their control quickly through most provinces (Lee 1970:174). Often political and social power was only surrendered by the local non-Manchu population because of the ability of the Manchu and Manchu-sympathizing gentry to protect them from marauding bandits or outside invaders.

The system of gentry control over local villages, and subservience of the gentry to a larger power at the provincial and national level, never worked with great efficiency even during the best of times in Qing China (Brook 1990:27–29). Corruption was the norm within gentry-controlled areas, and officials were easily bribed for favors by powerful non-scholar-officials or by outside influence (Hsiao 1960:166). This corruption also included the purchasing of Confucianist degrees by the merchant class, who for a long time were excluded from this process during the pre-Manchu period (Davis 1971:30; Smith 1994:81). Even though it tried mightily to institute a strong national government, the Qing Dynasty was never able to control fully all the workings of such a far-flung geographical empire as China (Davis 1971:14). This reality allowed the local gentry to exert more control over local areas, although they were most likely Manchu themselves or Manchu sympathizers. This disconnect between the national government and the local level, and ultimately the local and national governments' lack of connection to the population, fostered discontent from which secret societies grew stronger.

The rise of the Qing Dynasty correlated closely with the increasing connections of outside influences on what had long been a self-sufficient

Chinese society. For hundreds of years the economic system of China was largely internally self-sufficient and did not require a substantial trading network outside its controlled areas for sustainability. This system changed during the 18th and 19th centuries through the increasing influence of European powers along the coast. Ultimately the defiant adherence to little or no trade with outside entities instigated a series of wars, colloquially called the Opium Wars (1839–1842; 1856–1860). The battles between the traditional Chinese military and the technologically advanced European powers were heavily lopsided affairs, with the British and French securing rights to free trade in numerous ports and securing Hong Kong as a colony (Davis 1971:55; Spence 1990:158–159).

The Opium Wars and the abject failure of the Chinese military to defend sovereignty shook the nation. For hundreds of years the Confucian doctrine of non-trade with outside powers had placed the country in a technological vacuum, and ultimately led to the stagnation of growth in various sectors, including military advancement. This failure, while blamed partly on Confucianism, was mostly centered on the largely despised Qing Dynasty and its Manchu leaders. In the following years after the Opium Wars, peasantry and some local gentry, often bonded through secret society organization, attempted to overthrow their Manchu leadership through outright rebellion (Hsiao 1960: 472:473). The Taiping (1850–1864) and the Boxer (1899–1901) rebellions were a direct and painful result of the failure of the Qing to resist outside foreign influences.

Nearly fifteen years of bloody battles in the Taiping Rebellion dislocated hundreds of thousands in the rural peasantry from their lands in the war zone along the Yangtze River Basin (mainly in Guizhou and Yunnan provinces), and pushed them southward and toward the coastal regions (Davis 1971:31; Spence 1990:210). Resettlement of this population in already densely populated areas substantially altered the landscape of economic sustainability within these communities. For hundreds of years the rural agrarian communities had been largely self-sufficient due to the cultivation and consumption of crops and other necessary goods. Increased population density, through both war refugees and natural increases in population, overpowered the traditional system of self-sufficiency (Ownby 1996:17). Population growth during the high Qing times (1644–1800) reflected good government and technological advances (Ping-ti 1959:266–267), and the population of China nearly doubled during the eighteenth century (Spence 1990:93; Evans 1998:74). Agricultural land per person declined continuously across China from the 11th century forward, due to increasing population and land fragmentation (Zhao 1986:92). By 1833 the land per family had declined to less than three-quarters of an acre, for those farmers who could even find land. Ping-ti (1959:274) believed that nearly all the problems within the late Qing dynasty were the direct result of the increasing disparity between population and economic means (see also Elvin 1973). Increasing population density and decreasing agricultural lands disrupted the millennia-old systems at work in the country, and farmers soon became tenants on larger farms owned by the local wealthy elite (Hsiao 1960:386). This resulted, for the first time in Chinese history, in a substantial portion of the farming community relying on wage labor for their existence without any chance of owning property of their own.

The growth of coastal ports seized by the European powers during the Opium Wars represented another response to this massive internal migration of populations. Those dispossessed of land because of increasing population density in the farming regions, disbanded military men, orphans of war, and hundreds of other working-class individuals moved toward the ports because they offered opportunities to make a living. Due to the European influence, some of the major port cities Guangzhou (or Canton), Xiamen (or Amoy), Foochow, Ningpo, Shanghai, and Hong Kong had begun to industrialize for the production of materials to be traded both internally and externally (Spence 1990:210). This influenced tens of thousands of individuals who had once lived in the rural regions to migrate to urban areas for employment purposes (Davis 1971:32). According to Davis (1971:33), industrialization formed "a new urban working class [which] was created with the arrival of European manufacturing and commercial enterprises in the treaty ports. This class of wage-earning, migrant labour had no precedent in traditional society, integral as they were to a capitalist economy." As the port cities quickly became manufacturing centers, the rural farmers became more dependent upon wag -labor and craft production, which further altered the traditional Chinese society of self-sufficiency (Wong 1997:40)

In the haven of these ports controlled almost entirely by distant colonial powers, the long-strangled religions of Buddhism and Taoism found new life, as they were not as closely monitored by the Qing regime. In addition, the European powers did far more than merely provide employment opportunities. For example, close contact between the Chinese and their foreign neighbors created new problems for the traditional society, including the widespread introduction in port towns of Christian missions (which soon spread to the countryside) and the growing influence of Western political and social concepts upon all the classes of the Chinese population.

With the growth of these new social and economic realms, individuals long since excluded from the gentry in traditional Chinese society—the merchants—began to rise in importance due to their strong ties to the external colonial powers and the wealth that those ties brought them (see also Zelin 1990). No longer did the examinations and intellectual cultivation of individuals dictate positions of power within local and national government. Rather, wealth became the primary means of social mobility. According to Davis (1971:53), "This upsetting of the delicate equilibrium between wealth and status in determining the influence which an individual could exert in society damaged the hegemony of the bureaucratic ruling class beyond repair." Before the 19th century the merchants lacked upward mobility and power through both a decentralized production center in rural communities and the effects of the government's policies (Deng 1999:85). The movement toward emphasis on wealth as a precursor to power, rather than an adherence to Confucianism, allowed even the most common laborer and farmer to aspire to political goals through business ventures and hard work (Davis 1971:32–33).

In less than 300 years, the traditional Chinese society governed by Confucianism had fallen to an outside invasion by the Manchus; spread rampant corruption throughout the empire, fostering a strong resentment of the non-Manchu peasantry toward the leadership; and bent relatively easily to the forceful will of external colonial powers. The result of all of these events created a milieu of opportunity for secret societies to gain power and influence throughout China and to increase their membership without intensive effort.

In most cases, detailed scholarly study of secret societies is limited due to the relative paucity of documentary materials about the topic. After all, these were *secret* societies, and documentary sources tend to come from official records that are inherently biased (Ownby 1996:22). Due to the difficulty of obtaining clear records, it is nearly impossible to construct a list of secret societies in Qing China; it appears that societies had numerous and overlapping names, yet the Triads seem to have been the largest and most important secret organization around the world. While some secret societies existed in China well before the Qing Dynasty began in

1644, upheavals in this period in Chinese history created a mission for the societies and resulted in an increased number of organizations and growing membership. The new overt political goal of these secret societies was to overthrow the Qing Dynasty and restore the Ming (Lee 1970:17), and to do that, the societies had to increase their ability to organize the population. Secret societies were involved in both legal and illegal enterprises during this period, as even illegal activities helped to secure their goal of overthrowing the Qing Dynasty. As discussed in the preceding section, the historical context of the Chinese state from 1644 to the mid-1900s created an incredible opportunity for enlistment of new members and the spread of the power of these societies.

Several factors dictated the nature of a Chinese secret society, and these characteristics are quite comparable to those of other organizations around the world. Davis (1971:1–2) outlines the characteristics as follows: (1) voluntary, where membership within the secret society is based on a voluntary choice of the individual and not on the economic position, social status, or gender of the individual; (2) secret knowledge, where a secret society is based on the lack of information provided to the outside world about topics like the very existence of the group, its rituals and origin stories, and especially the current and past activities of its members; (3) secret existence, broken down into two types of societies—(a) where members do not know other members, or (b) where the members of the group are internally identifiable but secrecy exists in its connection to the outside world; (4) live and act together, members sometimes living communally or sharing property, a trait that can only work if there is a consenting basis of existence whereby all members choose to support the others; (5) dependent on members and non-members, meaning that a secret society must have the outside world to pit itself against, or there would be no reason for the association. Smith (1994:99) puts this succinctly: "In general, Chinese secret societies may be defined as associations whose policies were characterized by some form of religious, political, or social dissent from the established order."

It is important to note here that there is a distinct difference between the various Chinese secret societies, depending on the geographical location. In the north, secret societies such as the White Lotus were extremely religious in orientation, through both rituals and beliefs. In the southern parts of China, secret societies expressed religious values through rituals only but not in the organization or beliefs of the group as a whole (Smith

1994:98). This religious split between the groups, even though they were related movements, is important as it directly affected how they operated in the rapidly changing world. Specifically, southern secret societies were very appealing, as indoctrinated members did not have to alter their religious beliefs radically to join, and the benefits and aims were much more pragmatic than in the religiously based organizations in the north. In particular, the "southern societies derived their [origin] from the secular *pai-meng* or mutual-aid or friendship association" (Davis 1971:47). The majority of Chinese immigrants who spread throughout the world during the second half of the 19th century came from southern China (mainly Guangdong, Guangxi, and Fujian provinces).

In the rural areas of southern China, secret societies had existed for hundreds of years, arising out of the need to support the corrupt local governmental officials who controlled those regions. In many cases the scholar-officials who controlled villages or cities relied exclusively on the local populace to form militias during times of invasion or banditry (Hsiao 1960:294–295). As the control of the central government declined during the Qing regime, ultimately to the point of almost complete dissolution after the China was defeated in the Opium Wars, local peasants increasingly relied on established secret societies for protection, trade, and organization (Davis 1971:56; Wakeman 1972:33). Before the massive dislocation wrought by warfare, drought, famine, and population growth, secret society membership had remained small. With the massive internal displacement of people from established economic production areas to new regions with decreasing national control, and after the defeats of the first Opium War, secret membership among farmer peasants increased exponentially (Wakeman 1972:34; Ownby 1996:22).

Throughout the entire history of the southern secret societies, the main political goal was to overthrow the corrupt and oppressive national government, even before the Qing Dynasty (Davis 1971:73). These secret societies viewed the Qing Dynasty as the worst rulers in Chinese history, responsible for increased corruption and the failure of the Chinese military to defend itself (Spence 1990:169). In addition many members of the peasantry, who were forced off their land due to starvation or merely in search of better economic opportunities in major port cities, wanted to join these societies to throw off the shackles of foreign rule, both Manchu and European. According to Jean Chesneaux (1972:19), "the societies supported and led the struggles of the poor peasantry, the proto-proletariat (notably transportation workers), and other have-nots against the well-off and their political adversaries." In this way the Taiping Rebellion during the mid-19th century had deep connections to secret society leadership and organization (Stirling and Ward 1925:8). In particular, the secret society known as the Triads (also known as the Hung Men, Red Turbans, etc.), centered in South China, was extremely powerful in organizing the Taiping Rebellion.

The increasingly large wage-earning class in the port cities had completely changed China's social system. For the first time there were large numbers of people separated from ancestral agricultural lands, and families needed the support of secret societies to survive in the rapidly modernizing world. Family lineage was especially important in traditional Chinese society (Ch'u 1962:179–180), as evidenced by the perpetual control scholar-officials exerted over lands. In rural and urban settings, family and clan lineages had been the main glue holding communities together and organizing their production and livelihoods (Smith 1994:87, 91). Due to the ravages of war and famine, there was an ever-increasing number of individuals both in the countryside and in urban settings who had little or no family lineage left, and this was where the secret societies stepped in to assist as "substitute lineages" (Ownby and Heidhues 1993:4, 43).

The nature of the relationship between secret societies and family lineages is interesting in its ability to alter fundamental relationships without major conflict. Both these organizational schemes relied on the individual's sole adherence to that system (Davis 1971:82). Disenfranchised family lineages of the countryside did not necessarily support the scholar-official lineages controlling local politics and economics. In essence the only real conflict between secret societies and family lineages was in the opposition to gentry/scholar-officials, where the secret societies attempted to circumvent the control by these powers while local families might be part of, or in support of, these governmental figures. In other circumstances, the secret societies actually cross-cut lineage lines and banded together warring lineages, and even sometimes rectified class-based breakages within a single lineage, as the organization focused attention on the singular goal of overthrowing the government (Davis 1971: 81). This successful bridging of lineages gave secret societies significant power, except in areas where the gentry's lineage was strongly aligned with national military forces (Davis 1971:83).

Although secret societies explicitly avoided including the gentry in their fold due to gentry connections to the Qing government, sometimes gentry individuals and lineages did join, though only in limited and unimpressive numbers. These members had no extra power or entitlement to position within these organizations (Davis 1971:88). Exclusion of powerful lineages was the direct result of the Qing Dynasty policy of outlawing the societies, thereby effectively severing the possibility that a respectable (or aspiring) lineage might attempt to join a secret organization. Davis (1971:89) suggests that if the Qing had not placed the societies in an illegal light, these Chinese organizations might have followed an evolutionary path closer to that of European Freemasonry, with rich and powerful members comprising the leadership. The comparison and linkages of Freemasonry and Chinese secret societies are discussed later.

By the mid-1800s Chinese secret societies in the south were able to bring together an extremely variable group of both lineated families and isolated individuals, including dispossessed peasants, unemployed artisans, small traders, owner-managers of various land and water transport vehicles, general laborers, porters, and disbanded soldiers in proportions dictated by their location (Davis 1971:90; Ownby and Heidhues 1993:52). Depending on the region of the society's operation, the membership shifted toward domination by a single group, such as stevedores or transportation managers along the rivers; but nearly all came from a poor background, and all were part of the new working class. In some cases women had their own chapters of secret societies; their numbers within the larger groups are hard to quantify but did include those working as prostitutes as well as seamstresses and other laborers (Davis 1971:53).

Political goals alone did not entice this broad slice of Chinese society to join; instead there were pragmatic reasons for joining a secret organization. Economic connections provided through the membership of a secret society were probably the primary reason most individuals joined. The role of the secret society as a substitute lineage was reified in the recurrent use of terms such as *family* and *brothers* in official rules and rituals (Schlegel 1866:xii; Stirling and Ward 1925:24, 66–68, 82; Ownby and Heidhues 1993:84), and this bond was extremely important to ensure continuity with the traditional Chinese value of the family. Secret societies were made up of individuals spread across various economic spheres, and merchants and laborers alike could find trading partners even in largely unfriendly regions. For example, merchants traveling up the Yangtze River into territory held by warring or inhospitable gentry lineages could still find a means of organizing a trading opportunity through secret society memberships existing in the area (Wakeman 1972:29).

Relations between different secret societies (five or more separate main organizations and dozens of subchapters in south China) were not always cordial. In areas where a single secret society might have claimed control of legal pursuits such as river trade, or illegal pursuits such as smuggling, movement of another society into its realm might spark violence (Lee 1970:178). This led to the direct association of gang warfare with secret society activity throughout the government and among the non-affiliated populace (Davis 1971:79). In the early 20th century, rival secret societies fought for control of illegal activities across the United States, leading to several deadly incidents in San Francisco (Lyman 1974:491).

Membership in a secret society was largely a voluntary choice. In the cities, faced with innumerable individuals separated from family lineage ties, the societies found a large proportion of their membership willing to join without argument. In areas where the secret society sought to establish its control and will over a village or market, sometimes the collective membership of the society would coerce an individual or family to join through blackmail, threats of physical violence, and sometimes actual violence, including kidnapping (Davis 1971:88–89). Once having joined, an individual was a member for life, as the society could not take the chance that people might relate their experiences to an outsider or the government (Schlegel 1866:6). If a member did attempt to leave the organization, penalties could include beatings or perhaps even death. When joining a secret society the individual accepted a number of rules, each particular to the organization, that would secure group cohesion and solidarity. Prohibitions against leaving a society, divulging society information, turning in a fellow member to the law, and committing adultery with another member's spouse or daughter were all part of the agreement to which one adhered when joining (Davis 1971:90–91). Due to the secret and illegal nature of the organizations, penalties for breaking these laws were much more stringent than for breaking laws associated with the host society (Davis 1971:89).

Leadership of a traditional Chinese secret society in south China was based on a rotational system, so that every member had an equal opportunity to lead the organization for a set period of time. This rotation scheme decreased the chance that an organization would change from its original purpose of opposing the government, but it also decreased administrative costs, thereby decreasing the chances of the entire society being discovered by governmental officials (Davis 1971:43). In addition, there was no central leadership for the various chapters of a secret society, which was both a strength and a weakness in terms of the overall goals of the organizations. First, having no central leadership allowed each chapter of the society to pursue its own particular goals in a local environment without being controlled by distant leaders separated from the context. This allowed individuals to feel closely affiliated with the local membership, without fear of being subservient to other organizations spread throughout the country. The decentralized system also severely decreased the chances that an individual arrested or detained by the government could identify all the members or even the leaders of the entire organization (Schlegel 1866:6). A decentralized system of small isolated chapters was not easily organized for revolt (Wakeman 1972:31). This decentralization spelled doom to the Taiping Rebellion, as secret societies were a major player in recruiting soldiers and organizing the battles.

Chinese secret societies were able to capitalize on the growing wage-earning working class emerging after the Opium Wars, making themselves a truly dangerous political opponent to the Qing Dynasty. By performing a role as a "substitute lineage," the societies were able to provide opportunities, both legal and illegal, for thousands of Chinese at home and abroad. Membership in these societies allowed an individual to gain access to business contacts but also to form a kinship with others of a similar economic background. Secret societies effectively banded together widely disparate groups that earlier political regimes had pushed to the fringes. On a small scale these societies were effectively able to control a substantial portion of the economic activity in numerous areas across China, creating a new type of social organization that allowed for a large role in the eventual overthrow of an entire regime.

In sum, traditionally these societies served solely as mutual-aid organizations, but they morphed into politically motivated systems of rebellion and resistance to the Manchu national government during the 18th and 19th centuries. In resistance to the Manchu-dominated governmental system of the Qing dynasty (1644–1911 A.D.), these secret societies brought together individuals of the lowest social classes, and thus least powerful elements of Chinese society, into semi-organized religious and political entities spread throughout both north and south China. It is significant that secret societies were not so much anti-government as simply anti-Manchu. Historical analysts of the Qing Dynasty divide its history into two parts: high Qing (1644–ca. 1800), exemplified by peace, tranquility, and growth; and low Qing (ca. 1800–1911), identified by dynasty decline through social unrest, warfare, and external influence.

By the mid-19th century secret societies, outlawed by the Qing government, became a means of underground social organization in communities feeling the worst sting of the transition from a traditional economic system to proto-industrial core communities partly controlled by foreign powers. According to David Ownby (1996:2), in traditional Chinese culture "secret societies were informal, popular institutions, created by marginalized men seeking mutual protection and mutual aid in a dangerous and competitive society." Movement to proto-industrial centers stemmed from years of social unrest and changing rural economic systems, leading to the migration of dispossessed peasants. Dislocation of people from the land, and the destruction of self-sufficiency among thousands of families due to this process, created a substantial pool of unemployed or underemployed working individuals who had no access to social or economic prosperity within the Chinese political system without joining these secret societies. In essence, secret societies formed bands that crossed family lineages and political lines and provided an opportunity for merchants and laborers alike to increase their chances of finding employment opportunities, trade markets, or even political success through the accumulation of wealth.

At the same time as secret society membership and power within traditional Chinese culture grew, thousands of immigrants left the nation in search of economic opportunities across the world, including but not exclusively in Australia, Malaysia, New Zealand, the Philippines, and the United States. Overseas Chinese research has largely ignored the role that secret societies might have played in social organization within these host communities. Indeed, even today these secret societies dating back hundreds or thousands of years are affecting the global system of illegal human and drug trafficking, through semi-organized crime syndicates that still fall under the same banner as those earlier organizations. Understanding and researching the role of secret societies in historical contexts, including the examples from the state of Montana, will help scholars understand the dynamic nature of these institutions and how they have, and have not, changed during the intervening years.

### Chinese Secret Societies in North America and Montana

During the 19th century tens of thousands of Chinese voluntarily—if sometimes under pressing conditions—immigrated to countries around the world in search of opportunity. These immigrants came almost exclusively from the provinces of Guangdong, Fujian, and Guangxi, in the heart of the south China secret society territories. The role of secret societies in the United States is as complicated as within China itself. In China these secret societies were defined by their opposition to the Qing government, though they had other roles, as described earlier. As Chinese immigrants spread to other countries and farther away from the restrictive government, the role of secret societies underwent changes. As noted by Lee (1970:178):

*Gradually, the original revolutionary nature was lost and the groups became more in the nature of fraternities and mutual help organizations, although the secrecy about membership, rituals, and other details remained. The changed nature of the secret societies was particularly noticeable among the overseas Chinese in Southeast Asia. The exigencies of living abroad among the other people apparently helped to cement the ties of the Chinese, and in tying the Chinese together, the role of the secret societies was very important.*

For the first time in two centuries, these secret societies were able to work with almost complete immunity from governmental control and repression, and the true depths of their control over Overseas Chinese populations remains misunderstood. Shaoqing (2002:44–45) states that the Hung League, one of the largest secret societies, was far more significant in organizing the Australian Chinese population than scholars and the public currently understand. The Qing attempted to control some Overseas Chinese colonies through establishing consulates and even selling honorary titles to maintain loyalty (Spence 1990:211), although the Qing did not establish the first Chinese consulate in the United States until the late 1870s (Hansen 2006:42). These efforts met with limited success and never exerted much direct control over American Chinese immigrants.

By 1880 more than 100,000 Chinese men and more than 3,000 women resided in the United States (Spence 1990:213). Before 1862 there were six separate organizations that handled the immigration and production of Chinese migrants, and these merged in 1882 to form the Chinese Consolidated Benevolent Association, or Six Companies (Pan 1994:54). The Six

Companies, with direct ties to secret societies in China through similar legitimate businesses, organized substantial numbers of Overseas Chinese (Spence 1990:213; Pan 1994:54). As part of their efforts in assisting newly arrived immigrants, the Six Companies kept a detailed ledger on the identity, occupation, and location of the Chinese members of their organization (Hansen 2006:41). This registry allowed the Six Companies to organize immigrants into collective labor efforts in their current location or transfer them to where more labor was needed (Hansen 2006:41). Baxter (2008:30–31) notes that the Chinese Six Companies represented a direct means of using collective power to resist the influence of anti-Chinese agitation during the 19th century. For instance, after the Exclusion Act of 1882 the Six Companies facilitated the illegal immigration of Chinese workers by the thousands across the Canadian and Mexican border. The Six Companies went even further by providing the illegal immigrants with falsified documents of residency and helped to construct a new identity for them, including family relationships, occupancy, and educational background to assist in the avoidance of American law enforcement (Hansen 2006:51–52).

Culin (1890:2–5) thoroughly discusses the role and extent of Chinese secret societies in the United States, noting the presence of Chinese secret society chapters in Chicago, St. Louis, New York City, and Philadelphia, with a large headquarters in San Francisco. According to some figures, upward of 80% of Chinese living in Tucson, Arizona, during the 1880s belonged to the Triad secret society, which was deeply entwined with the Six Companies (Dirlick and Yeung 2001:xxiii). Pioneering work by Paciotti (2005) determined that the significantly higher murder rate among the Chinese in Seattle between 1900 and 1940 was a direct result of secret society organization and the intense warfare that disputes brought to the community. Within the Chinese Six Companies there was continual internal strife, and several splinter groups formed during the 1890s and early 1900s (Hansen 2006:40). Even in the early 21st century there are internal tensions within the Six Companies, more commonly known today as the Chinese Consolidated Benevolent Association.

In Prescott, Arizona, the Chih-Kung T'ang, also known as the Chee Kung Tong or Triad Society, operated as a means to protect the lives and businesses of Chinese in that community, with a membership percentage similar to that of Tucson (Lister and Lister 1989:51–52). Based out of San Francisco, it had branches in Nevada, which included mercantile operations (Hunt-Jones 2006:36). According to Hansen (2006:49), the Chee Kung Tong, or a variant, was critically involved in human trafficking cross the U.S. and Mexico borders during the early 1900s. In addition Cheung (2002) discusses how ritual and performance within the Chih-Kung T'ang meeting hall in Tucson acted as a means of resistance to the world of exclusion and racism existing in the United States during the 19th century. Outside the meeting hall the Chinese of Tucson and other areas were beset on all sides by negative attitudes and laws, but within the Chih-Kung T'ang the reenacted dramatic episodes of warfare and symbolism empowered the participants (Cheung 2002:42, 46). The analysis of Chinese in Arizona by Lister and Lister (1989), Dirlick and Yeung (2001), and Cheung (2002) demonstrate the powerful role secret societies played in Overseas Chinese communities.

In the early 1960s Lyman and colleagues (1964) discovered an assemblage of Chinese secret society trappings while doing research in the Cariboo region of British Columbia. These trappings included a ceremonial robe, books of codes, and a long board with Chinese calligraphy. When translated, the calligraphy was a list of rules for the Chih-Kung T'ang or Triad Society, a Chinese secret society operating in California before the mid-1850s, with a British Columbia chapter that opened in 1862 (Lyman et al. 1964:530–531). According to the translated rules, the Chinese in British Columbia used secret societies for mutual aid, protection, and other social activities (Lyman et al. 1964:532). Far from the stereotypical image of secret societies as havens for crime, the rules of the Cariboo Chinese Chih-Kung T'ang included the reason for founding the society: to "maintain a friendly relationship among our countrymen and to accumulate wealth through proper business methods for the benefit of all members" (Lyman et al. 1964:534). The documents also outline the rules of the society's activities, including promoting fair business practices among members, prohibitions against in-fighting during visits to brothels or gambling houses, hospitable housing practices, and a number of other rules backed by declarations of punishment (Lyman et al. 1964:535–539). Overall, the discovery of the secret society materials in British Columbia shows the widespread power of these organizations and their ability to conduct both formal and informal business and social maintenance through their activities.

The largest organization, which also had direct links to Chinese secret societies, was the Six Companies based out of San Francisco, California.

This group was probably the single most important immigration and social organizing force for Overseas Chinese communities in the United States and Canada (Dirlick and Yeung 2001:xx); a similar situation may apply in other countries. The Six Companies, an organization of entrepreneurs at the top, often paid the expense of travel for those wishing to immigrate; upon arrival in America immigrants would pay off their debt to the company through their labors. During this time the Six Companies would feed and protect these new immigrants, for a fee (Hunt-Jones 2006:36). Chinese employed by the Six Companies—or perhaps indentured is a better term—would be sent to various locations where their labor was desperately needed; such locations were mainly railroad or mining ventures in the American West (Sadowski-Smith 2008:783–784). While this organization was definitely not a secret body, its extremely weak links, if any, to the Chinese government suggested that a powerful autonomous secret society was pulling substantial strings. Even long-distance trade in imported Chinese goods was channeled through the extensive network of the Six Companies, a reality that likely exerted yet more control over an individual's life (Dirlick and Yeung 2001:xxiii). While these societies in the United States did move away from revolutionary tendencies, at least overtly, and toward mutual aid and social organizational roles in which they did not attempt as significantly to hide their clandestine activities, they were still connected fundamentally to their roots in China.

Although Swartout (1988) discussed the history of the Chinese in Montana, his overview displays a somewhat inaccurate portrayal of the immigrants' social organization. Swartout (1988:48) states that these immigrants "were able to maintain or to create new social organizations based along traditional clan or district lines." This is in part true, as these affiliations did have manifestations in Montana, yet secret societies created social and economic linkages among all members of the Chinese community, regardless of district or lineage. In Montana there are quite a few clear examples of the direct impact of secret societies, including the Six Companies, upon both the Chinese and non-Chinese communities. In 1873 the Six Companies in San Francisco issued an order to the local head of that organization in Pioneer, Montana, to buy as much land as possible. This was likely ordered to initiate a dispute over the recently passed Montana Alien Law forbidding Chinese from owning land or mining claims (*New Northwest* 1873a:4). The local newspaper editors felt that this effort would ultimately fail, as it would quickly be tested in court if they attempted this action. It is unclear how successful this attempt at mass claim purchase was, but court cases in 1874 overturned the law, as discussed in chapter 2 and Wunder (1980).

Even as early as 1867 the presence of these companies was detailed in Montana with the death of "China Mary." As noted in chapter 1, China Mary was a prostitute in Helena who was murdered by unknown parties just before her departure to California to reunite with her children and marry an African American who had worked for her in Helena. According to newspaper accounts, Mary had garnered the animosity of other Chinese in Helena by denouncing her citizenship and refusing to "conform to their 'company' arrangements" (*Helena Daily Herald* 1867a:2). As a result of this conflict, only four other Chinese prostitutes joined her funeral procession. Denial of proper funeral rites was a significant penalty to the Chinese immigrant, due to the religious necessity of such mourning (Chung and Wegars 2005). As China Mary was a prostitute, and a Chinese one at that, there was little investigation into suspects or motives for the murder, and the case was never resolved. It is unclear whether Mary's dispute with the "company arrangements" was strong enough to illicit a murderous response, but it is clear through other stories that these organizations did exert both coercive and sometimes deadly power over Chinese living in Montana, as further discussed later.

The Six Companies also served a regulating function for the immigrants, as demonstrated in the life and death of an elderly Chinese man in Park City, Montana. According to the newspaper reports, the deceased person had been exiled from China and, once in United States, had been harassed by the Six Companies. It seems as though this individual, who never shows up in any subsequent reports, had somehow alienated himself from both the government in China and the powerful Six Companies in the United States (*Helena Independent* 1882f:3). This small article suggests that the Six Companies exerted significant control over individuals' lives and helped to shape the social organization of Chinese communities in places like Montana.

A different example illustrates that the Six Companies had control over women sold into prostitution and slavery. In 1884 Chin Cum's contract with another Chinese woman, Suzie Mouie, had expired, but Chin Cum was still held against her will. A trial over the affair apparently used a Chinese interpreter with clear ties to the Six Companies, and the case went against Chin Cum. She was placed in protective custody in the Lewis and

Clark County Jail until such time as the court could find a suitable home for the woman to serve as a domestic servant (*Helena Independent* 1884d:3). While this story ended positively, it indicates the power of the Six Companies as deeply connected to prostitution and to the treatment of Chinese women as property or slaves. In a similar case, a woman betrothed in China was abducted by mysterious parties and sent to work as a prostitute in California; a "wealthy Chinese company," most likely the Six Companies, "owned" her. The husband followed the woman to California, but the company sent her to Montana, where finally she was overtaken by her husband and rescued. Numerous kidnappings and counter-kidnappings followed over the next three years, until finally in 1881 they were married according to United States law and housed in the Lewis and Clark jail for protection from the angered company (*Helena Independent* 1881d:3).

Not only did the Six Companies exert control over the population, but other more local secret societies seemed to have a certain amount of influence over Montana's Chinese communities. Far away from Chinese governmental control, such organizations were not shy in displaying this coercive power and were largely ignored by the United States and Montana governments, since the events associated with such control were treated as internal Chinese matters. Take, for example, the story of Gem (or Jim), a Chinese laundry operator in the town of Deer Lodge. According to newspaper accounts, Gem was approached by the Chinese Gambling Society at Deer Lodge for recruitment, and upon his refusing this offer, the society "decreed his assassination" (*New Northwest* 1874d:3). Gem appealed to the local non-Chinese community, including the newspaper, for assistance, and they ran editorials denouncing the threats (*New Northwest*,1874e:3). Finally an apology letter written by Gem to the Hung Seung Tong Company, which probably had ties with national and international societies, appeared in the August 8, 1874, edition of the *New Northwest* (1874f:3). Several months later on March 12, 1875, the *New Northwest* received a letter from Gem, who had left for his home in China; the letter arrived shortly after Gem's death on board a steamer that sank off the coast of China. It is unclear why Gem had left Deer Lodge, but it seems logical to assume that the difficulties with the Hung Seung Tong Company did not end with an official apology letter and might have precipitated his departure. It was likely sheer coincidence that the steamer Gem took back to China sank before final arrival. In a similar case Sam Yek, another Chinese laundry operator, disappeared after a dispute with the Nine Companies, which was yet another indistinct secret society, and editors supposed that the company might have been involved in some manner (*Helena Independent* 1882e:3).

There appears to have been a centrality to secret society efforts in Virginia City, Montana, one of the first areas of gold discovery in the state. This was one of the original points of immigration into Montana for the first wave of Chinese. According to numerous resources there was a Chinese Masonic Temple in Virginia City, dating to at least the 1870s, that "stood on a little promontory near the town's entrance" and "loomed high in the skyline, an oriental outpost on the borders of western barbarism" (*Mineral Independent* 1941:4). Excavations by the University of Montana in 2000 were near the location of the Chinese temple lodge (Bowen 2004), though crews did not locate firm evidence of this significant structure. Due to the 1950s road realignment through Virginia City, there are likely few remaining subsurface deposits associated with this structure.

In many cases the host communities commonly called these Chinese secret societies, particularly the Triad, the Chinese Freemasons. In Overseas Chinese communities there were dubious linkages between Freemasons and these organizations; outsiders called these institutions Freemasonry lodges because of their secretive nature. Even members of the Freemasons attempted to find linkages between Chinese secret societies and Freemasonry, with limited results besides some similar iconography and rituals (Glick and Sheng-hwa 1947; Stirling and Ward 1925). Other Freemasons however, have made the claim that even up through 1882 the Chinese did not have any Freemasonry connections, though they were starting to incorporate some traits (Giles 1882:379). In 19th-century Australia the Hung League altered their name to Chinese Masonic Society as a means of explaining their organization to non-Chinese and helped to foster some positive interactions with white Australians (Shaoqing 2002:42–43). Indeed, the excessive pomp and attendance at a funeral of a Chinese man in Helena was discussed in relation to his position as a head person among the Masons (*Helena Independent* 1884c:3).

In reality the Chinese Masonic Temple of Virginia City, Montana, was merely a district meeting house of the Triads, Six Companies, or another strong secret society. Over the following years it appears that this secret society exerted strong powers over personal freedoms, commerce, and social organization, not only in the immediate area but across the state, and it was "reputed to have been, at one time, western headquarters of Chinese Masonry in the United States" (*Mineral Independent* 1941:4). A tapestry from this Masonic Temple in Virginia City, dating to 1876, details the rules and regulations of this organization and includes the same types of language and rules as applied in secret societies in China, as discussed earlier. In particular, the banner had rules including not to "covet the wife or sisters of brethren," and required secrecy at all costs with a rule that members "may not compromise any other brother," with potential punishments of up to 306 strikes of a cane (Swartout 1988:50). Like similar rules in China, this banner created the means necessary to retain social cohesion within the secret organization and continued the use of the language of "brethren" and "brothers" as a substitute lineage.

In the late 1870s a major conflict between rival Chinese companies in Virginia City cost the lives of two individuals. Although its account was published many years after this conflict, the *Mineral Independent* provides some clues about the root of the disagreements related to this event and its connection to larger organizations:

> *Six Chinese companies operated in Alder gulch, and every Chinese in the diggings was affiliated with one or the other of them[;] whether the companies were connected with the six great Chinese tongs in control of the Chinese colony in San Francisco is not known, but the impression prevailed in Virginia City that the connection was direct. The head of each of these companies was martinet and his followers obeyed his orders implicitly. Their quarrels the head men settled among themselves. There were many differences over leases and gradually there grew up two factions in this Chinese fraternity house, four companies aligning themselves against two and among the white people these factions became known as the Four company and the Two company groups. (Mineral Independent 1941:4)*

In effect, during a conflict that lasted nearly a week, the warring factions who used the stage road as a common boundary cut off Virginia City from the rest of Montana. Eventually, after the death of two members of the Two Company, the conflict subsided, and the sheriff arrested two individuals for the murders. Chinese manipulation effected a mistrial, and when the defense asked for the bodies of the two murdered Chinese, officers excavated the graves, only to find them empty. Without corpses there was no case, and the accused individuals were released. There were other "tong wars" reported through secondary literature in Butte during the 1910s and 1920s (Lee 1978), though the scale and death rate seem overblown compared to primary historical resources (*Butte Miner* 1922:3).

In another situation the interconnections between Chinese societies in other cities and Virginia City became clear. A case in point included the events surrounding Han Bo, a Chinese Methodist convert who arrived in Virginia City to begin proselytizing the local community. It seems that the Han Bo was not a member of the local secret society, and this group, seeing its power possibly being challenged by outside influences, enlisted the help of a sister organization in Helena. The nature of the dispute illustrated the divide between Han Bo and the other Chinese in Virginia City:

*His enemies were determined to get rid of him in some manner, and the facts seem to show that the Helena branch of the secret order was instructed to procure his removal to this city upon a criminal charge. A complaint charging him with the [theft] was therefore made before Judge Davis, and last week Sheriff Jefferis went over to Virginia city, arrested Han Bo and brought him to Helena. (Helena Independent 1881a:3)*

Han Bo claimed to have never even been to Helena before the day he was arrested and taken there and that, on the day of his arrival in Helena, the individual who had put out the warrant "came to the jail and said he was not prosecuting the case because there was any truth in the charges, but because he was compelled to do so by the instructions of the secret order" (*Helena Independent* 1881a:3).

After the inconsistent testimony of those accusing Han Bo of the robbery, the judge dismissed the case in favor of Han Bo. While relishing the dismissal of the charges, Han Bo requested that he be housed in the jail for an indeterminate amount of time as he was worried that if he left that day, he would surely be murdered by agents of the secret society. The court granted his request, and the case never appeared again in the newspaper, suggesting that Han Bo survived after discharge from the Lewis and Clark County Jail. This account indicates that secret societies not only had power over the immediate local community but, through connections to other organizations, could control individual action to a somewhat significant extent.

Perhaps the most intriguing story in Montana Chinese history with apparent connections to secret society activity is the 1928 murder of Mow Lu Dun (or Mar Leu Dunn in some reports). Lu Dun's murder was a sensational story in January and February of 1928, though in the end no one was formally charged in the case. According to the reports, Mow Lu Dun, aged 55, was murdered at his garden plot southeast of the Green Meadow

Farm, which lay north of Helena's downtown (*Helena Independent* 1928a:4). Lu Dun had been accosted in his own residence, shot three times with a .32 caliber pistol, and had his tongue removed and throat slashed with a knife (*Helena Independent* 1928a:4, 1928b:3). Initial investigations by the Helena police located tracks in the snow outside the residence, suggesting that the murderer(s) had lain in wait until Lu Dun returned from town with money needed for paying the lease on the house and garden property (*Helena Independent* 1928a:4). The angle of the bullets in the body and door frame suggests that the murderer was inside the house and waited for Lu Dun behind or on the living room couch. Police found a .25 caliber automatic pistol on Lu Dun's body, but it was unfired (*Helena Independent* 1928b:3).

While the murder itself was sensational, the events that followed provide a clearer image of internal Chinese society during the late 1920s. Wong Chuck, a local Chinese drifter, was arrested the same night as Lu Dun's murder and held as the only suspect in the murder. According to police reports, Chuck's clothing was wet and covered in a variety of cockleburs found near the garden, though Wong claimed he never even left his hotel room that evening (*Helena Independent* 1928a:4, 1928b:3). In addition, Wong Chuck stated to investigators that he had heard of the murder at 10:00 a.m. on the morning of the crime, though the body was not discovered until 11:00 a.m. (*Helena Independent* 1928b:3). In the days following Chuck's arrest, investigators noticed that he had no Chinese visitors, something that was unusual enough to warrant a full-length editorial (*Helena Independent* 1928c:4), but investigators were still interviewing other suspects in Helena's Chinatown (*Helena Independent* 1928d:3).

Within a week of the murder the Chinese community in Helena began to withdraw from the investigation, and many changed their stories. Upon first interview, prominent Chinese individuals noted that on the day of the murder Mow Lu Dun withdrew $350–$400 from his account in town (*Helena Independent* 1928a:4, 1928e:4). By January 10 the same informants stated that the amount was only $40 (*Helena Independent* 1928e:4). While the Chinese community became more silent in their interviews with investigators, many also pulled away from helping to arrange the burial of Mow Lu Dun (*Helena Independent* 1928f:4). The first indication of the scale of events came on January 13, when Mow Lu Dun was buried at Forestvale Cemetery. Instead of the local community providing the funds for burial, a wealthy Chinese man from Seattle, Mar Kin Dun, provided all costs for the funeral on his first day in town (*Helena Independent* 1928g:4).

On January 17 police released Wong Chuck due to the establishment of an alibi (*Helena Independent* 1928h:3), and in reaction to the protestations of Mar Kin Dun and another wealthy Chinese merchant from the Pacific Coast, Toy Young.

In addition, the case brought the attention of a certain R. Woodside, a European American man from San Francisco. According to the newspaper reports, Woodside was fluent in Chinese due to his long-term habitation in China and his marriage to a Chinese woman. More interesting, perhaps, was that Woodside was a member of Dr. Sun Yat Sen's revolutionary party in the 1911 overthrow of the Qing government (*Helena Independent* 1928i:3). Sun Yat Sen's use of secret societies, and other Overseas Chinese voluntary organizations, was the key facet in his plans to overthrow the regime in the 1910s (Liew 1985:283; Zelin 1990:94, n.32). Woodside stayed at the Placer hotel in Helena while he searched for the real perpetrator of the murder, under the pay of a Chinese agent in San Francisco. By late January 1928, Woodside and other investigators had found no additional suspects, and the case faded from the newspapers. The last known information on Woodside's involvement was in relation to the death of a European American oil worker who was thrown underneath the wheels of a locomotive in Great Falls. In the oil worker's possession was a handbag full of correspondence relating to Woodside and the Chinese murder case, including letters and depositions (*Helena Independent* 1928j:4).

The Wong Chuck connection to this story does not end with his release on January 17, 1928. After being released from the Lewis and Clark Jail, Chuck moved to Butte and by February was seen with completely new wardrobe suggesting sudden wealth. After Wong Chuck left town, some of the Chinese in that community began to leak information that he was indeed the murderer; that there were closed-door dealings that had precipitated the murder of Mow Lu Dun; and that Mar Kin Dun only came to Helena to secure his release from jail (*Helena Independent* 1928k:3). Chuck continued to interest authorities, as he was arrested in Butte during October 1928 for selling and using opium, which resulted in hefty fines (*Helena Independent* 1928m:4, 1928o:4, 1929a:4).

While Wong Chuck faded into the late 1920s, the story of Mow Lu Dun's murder remains a captivating tale of intrigue, murder, and possible secret society activities. The Helena Chinese community's silence in the case suggest an internal investigation and prosecution relating to the murder. These investigations seem associated with the arrival of Mar Kin

Dun, Toy Young, and R. Woodside, though their true involvement remains forever shrouded in mystery. From all the evidence it appears that Mow Lu Dun garnered the wrath of unknown parties for providing information or speaking out of turn, as evidenced by the removal of the victim's tongue through his throat. The failure of the Helena Chinese to offer testimony also attests to some type of strong linkage between Wong Chuck and an internally organized apparatus. Finally, Wong Chuck's release and subsequent acquisition of new and expensive clothes in Butte suggest a successful conspiracy and a reward for the task completed. The full extent of the story may never be discovered, but there was far more strength in internal Chinese social organization than is often discussed, even in the final historical period of Chinese residency in Montana.

It is clear through these newspaper accounts that Montana, among other states, had active and powerful Chinese secret societies at work within their borders. These organizations controlled the movement of labor, prostitutes, goods, and even non-members through coercive and sometimes violent methods. Secret societies even played a role in the burial of the dead in the Mow Lu Dun murder case in Helena and in the death of an aged Chinese gardener in Missoula (*Anaconda Standard* 1907:7). The complete role of these societies remains unclear, but it is apparent that Chinese communities in the United States were shaped in part by institutions begun in China. It is also unclear how much of the wealth created in the United States made it back to China to help support insurrections led by these secret societies in the homeland. As demonstrated by Culin (1890), Lyman and co-workers (1964), Chen (2001), and Hansen (2006), many secret societies used their communal pool of funds to assist immigrants and also to fund various economic ventures. It is clear that many of the major purchases of Chinese in German Gulch and Blackfoot City, to name just two examples, used funds funneled through these voluntary organizations, whether they were the Six Companies or a less visible society.

By the 1940s most of the Chinese living in Montana had left, taking with them the powerful secret societies operating in the state. Areas along the coast of the United States continue to have substantial Chinese populations, namely Seattle, San Francisco, Portland, Los Angeles, and San Diego. Many of the Chinese across the United States resettled in these places as they left inland rural areas, and the strength of the secret societies continued in those larger urban centers.

## Anthropological and Archaeological Implications of Secret Societies in Montana

Tracing the history and current status of Chinese secret societies provides the context necessary to understand the role of these groups as forces of social organization. There are numerous means of understanding social organization within traditional societies (e.g., Boas 1897; Steward 1955; Geertz 1961; Warner 1963; McGuire 1992), yet there is no consensus, and each position has strengths and weakness too lengthy to discuss here. It appears quite clear that in traditional Chinese society the main unit of social organization was the family lineage, and this continues to be a strong force in China today (Woon 1984:105). Yet secret societies controlled and in some cases still control some aspects of social organization within the Overseas Chinese communities around the world. As described by Ownby (2001:153), secret society growth in China resulted from "efforts of young men, marginalized and indeed driven from the village by population pressure and commercialization, to arm themselves with protective structures in a world where local lineages no longer were able to do the job." Leadership in these secret societies required substantial financial influence and established trade networks, leading to the domination of these groups by Chinese merchants or wealthy entrepreneurs, as illustrated in British Columbia (Lyman et al. 1964; Chen 2001; Morrow 2009).

In Montana, secret society chapters linked directly to the Six Companies in San Francisco (or another society) and then in turn connected to sister/parent organizations in Hong Kong and mainland China. Newspaper accounts of secret society involvement in the daily affairs of Montana Chinese life during the 19th and early 20th centuries suggest that consanguineal family lineages were not as important as the substitute lineages provided by these voluntary organizations. Chinese secret societies in Montana heavily controlled trade of goods, movement of labor, and interactions with the host society and maintained internal social cohesion through both coercion of and retribution to those who failed to follow the organizational rules. For instance, Chinese secret societies were critical in the importation and distribution of smoking opium in the United States, before and after anti-opium legislation (Ahmad 2007:23). In some rural British Columbia mining communities, secret societies became the only social organization controlling significant parts of the community activity (Lyman 1974:485)

From an archaeological perspective, there is serious reason to take a more inclusive view of Chinese social relations during this period. In a recent thesis Trelle Morrow (2009) analyzes the use, distribution, and symbology of Chinese stoneware in overseas contexts. As discussed in this chapter, Morrow (2009:31) notes that many merchants in diaspora communities had close ties to secret societies and other voluntary associations. These ties were integral to establishing a complex network of trade that facilitated the importation and distribution of traditional Chinese goods. Specifically, Morrow (2009:86–87) states that the Overseas Chinese laboring classes represented a captive consumer market that relied on the merchant classes for the goods. In addition the merchants could dictate the market availability in terms of what type of goods and in what quantities (see also Hsu 2006). According to Lyman (1974:484), "so long as the Chinese were either denied the right to testify in public courts or unwilling to employ the American legal system, the merchant-elite of Chinatown exercised an awesome authority," which undoubtedly included the sale and distribution of merchandise. Chen (2001:246) notes that members of the Hung League, a globally powerful secret society sect, owned four of the five Chinese stores in Stanley, British Columbia, during the 1870s and 1880s. Stores, goods, merchants, and consumers were inextricably linked to secret society activity and influence, especially in the 19th and early 20th century.

In many Montana communities, rich Chinese merchants served as unofficial mayors (Rossillon 2008:15–17) but also helped to purchase and manage mining leases and claims. For instance, in the German Gulch Chinese community the owner of the Chinese store also held deeds on several of the mining claims in the district (Fredlund et al. 1991; Meyer 2001). Chinese working in German Gulch relied on the Chinese merchant not only for the goods of his store but also for the employment opportunities in the gulch. During its peak German Gulch sustained a population of several hundred Chinese laborers, which would have represented a profitable market for the local merchant. Within this framework it would be difficult to separate ethnic boundary maintenance as social action against the dominant European American population from the display and purchase of goods from the Chinese merchant for other reasons, as there was little other choice. This perspective does not limit the agency of a Chinese individual, though there were far more factors involved in purchasing a celadon bowl versus a European American version than previously thought. Thus any interpretations of acculturation or assimilation must be couched in understanding the basic availability of goods and the direct social impacts of not purchasing from the local Chinese provider.

Excavations by Hall and colleagues (2006) and Rossillon (2002, 2003, 2008) of properties and features occupied by the Chinese discovered a variety of Chinese and European American goods. In their view the ratio of Chinese to European American goods reflected the level of assimilation or acculturation. It seems more plausible that the artifact ratios reflect market availability more than assimilation and acculturation processes. As noted by Voss (2005:432), "it would be a grave ethnocentric error to assume that cultural change in Overseas Chinese communities was necessarily the result of acculturation to European American cultural norms." Isolated Chinese camps like those excavated in German, Cave, and China gulches had limited access to a variety of markets, but all possessed a Chinese store or direct trade with an extra-local source. In contrast, the mining camp on Emigrant Creek near Yellowstone National Park lacked any local Chinese markets, and it appears that occupants made do with locally purchased European American goods and the few curated Chinese wares they brought with them. The simple fact of market availability precluded the easy access to traditional goods and opened up more economic choices.

Perhaps the best example of merchant-class control of goods is the construction of the Central and Northern Pacific Railroads (NPRR) in the 1860s and 1880s. The remarkable similarity of artifacts found in all sites along both grades is a direct result of constrained contracts with merchandisers. Of the eight known Chinese sites along the NPRR in Montana, all exhibit the same ceramic ware types (Celadon, Bamboo, and to a lesser extent Four Flowers style) and the same array of opium smoking paraphernalia. In essence each stop along the grade, regardless of the location in Washington, Idaho, or Montana, was created from the same constituent parts, likely supplied by the same Chinese merchants. Chinese laborers had limited or no market access, except to the same merchants who negotiated their wages and work schedule. From the lack of any standing in China, the merchant class of Chinese immigrants in diaspora communities cornered significant power and influenced the lives of thousands.

While it is clear, and should be emphasized, that not all members of the Chinese community were members of these organizations, their experiences were nonetheless shaped in part by these societies. Voluntary organizations, whether they were clans, district associations, or secret societies, formed the basic unit of social organization in Chinese communities. For instance, in the battle against restrictive immigration legislation the Chinese Six Companies, a hybrid of benevolent (or district) organizations and secret society, was the main national organization rousing opposition to this law. In Montana the Chinese Six Companies organized local chapters against the state Alien Law and federal Geary Act. Regardless of an individual's membership in the Six Companies or an allied group, the efforts of these organizations affected the host community's perception of Chinese immigrants. Both the Chinese Six Companies and the Bao Huang Hui or Chinese Empire Reform Association served as mutual-aid and mutual protection societies in the 1800s and early 1900s.

As Baxter (2008:33–34) notes, the Chinese were not passive or docile actors in 19th- or 20th-century politics. Due to the widespread anti-Chinese violence sweeping the American West from the 1870s through 1890s, Chinese armed themselves and prepared for the worst. In the "Tong War" of Virginia City, Montana, the temple and fraternal lodge served as a central point of strategic planning and eventually peace talks. At German Gulch excavators located over 140 spent shell casings from pistols, shotguns, and rifles within only a few square meters (Fredlund et al. 1991; Meyer 2001). These shells clustered in and around the Chinese store established in German Gulch during the mid-1870s. While there are likely a variety of explanations for the number of shell casings within such a small area, including hunting, the presence of these artifacts in direct association with the archaeological remains of a Chinese store provide some interpretive potential.

If voluntary organizations like secret societies were the main means of offering collective resistance against both legislation and anti-Chinese mobs, could the shell casings from German Gulch represent an effort to prepare for the worst and display collective power? Excavations from China Gulch in Louiseville in western Montana recovered only a handful of shells and cartridges (Merritt 2007, 2009a) suggesting some disparity in the two rural Chinese mining assemblages that could potentially be accounted for by anti-Chinese agitation and collective response. Weisz's (2003) work along the NPRR discovered dozens of shell casings associated with the Chinese camps, suggesting that these workers were indeed well armed, though there was little need to hunt local game.

As scholars investigate the Overseas Chinese community anthropologically, archaeologically, geographically, or historically, the role of these secret societies must not be brushed aside as insignificant. Scholars have demonstrated the significance of voluntary organizations to immigrant populations (Little 1967; Kerri 1976; Brettell 2005), especially among the Overseas Chinese (Lyman et al. 1964; Lyman 1974; Lister and Lister 1986; Cheung 2002; Baxter 2008). It is difficult to gather substantial documentary or even oral history sources on these organizations because of their secretive nature, their ongoing existence and presence, and the potential adverse consequences for informants. If a researcher is aware of the importance of these organizations, and how they operate(d), their true role might be understood. At the center of Overseas Chinese experience is the search for economic opportunity, but the internal social structure of this community is largely secret societies in their role as substitute lineages. Merchants, due to their now privileged position at the top of the diaspora hierarchy, effectively used all modes of social organization to their own advantage while at the same time benefiting and restricting the other Chinese immigrants.

Merchants served as community figures in the 19th- and 20th-century Montana communities. These Chinese businessmen sponsored and capitalized mining ventures, and supported or influenced critical legal battles in Montana's courts, and their ever-present linkage with the new forms of social organization directly affected the material culture imported and used in the state. Archaeologists discover the material culture remnants of these social relationships, and simply noting the presence or absence of Chinese items as signs of acculturation or assimilation simplifies the social relations. Praetzellis and Praetzellis (2001) illustrate how a Chinese merchant in California may have used material culture as a symbol of tradition or assimilation based on the social setting. Wegars (1993), Voss and Allen (2008), and Voss (2005) push for archaeological research that focuses on deeper understandings of the transnational workings of Chinese society, instead of the site-specific discussions of particular historical circumstance. An archaeological framework using a firm understanding of Chinese social organization in China and in diaspora communities is necessary to create more articulated models of social cohesion and change. As noted by Mullins (2008:156), anti-Chinese sentiment in the United States and other diaspora countries deeply affected social relations both internally and externally, as seen by the growth and strength of voluntary organizations.

Beyond roles of secret societies and the merchant-class Chinese control of labor and opportunity, the comprehensive analysis of the Montana experience for these immigrants highlights many roles.

# 6. Conclusions and Future Directions

## The Chinese Experience in Montana

From their entrance into Montana's mining frontier during the early 1860s, the Chinese left indelible reminders of their contributions to the state's historical development. Facing seemingly insurmountable racism and legal exclusion, this population helped to spur the economic growth of Montana through gold mining, service-based industries, and constructing the Northern Pacific Railroad. Over the past six years the Chinese story has been pulled together from the historical and archaeological record in an effort to provide these immigrants with some acknowledgment for their achievements. Unlike California, where substantial descendant populations still live, Montana has only a few families who can trace their roots back to Chinese pioneers of the state. A descendant of a Philipsburg Chinese family, the Lees, returned to that community in 2006, the first revisit to the Montana community since their departure in the 1920s or 1930s. The Lee family owned and operated a restaurant in Philipsburg, and a family photo from 1910 hangs in the Granite County Museum in Philipsburg (Paige et al. 1998:9). As noted by Voss (2005), future Overseas Chinese archaeological projects must engage with descendant communities, but this is far more difficult, if not impossible, when the majority of that group no longer lives in the state under study. Until descendants come forward to share their stories, photographs, and other records, historical and archaeological resources are currently the best and only data sets for understanding the Chinese heritage of Montana.

The principal goal of the overall research effort was to consolidate primary and secondary historical resources to present a history of Montana's Chinese population. This objective relied heavily on the use of period newspaper accounts, census records, and to a lesser degree deeds, mine claims, and other primary records. Since Chinese perspectives are often absent from these accounts and records, it requires careful interpretation of these primary records to provide details on the trials, tribulations, and successes of this immigrant population. Secondary historical resources were often a disappointment due to their limited perspectives that generally presents the Chinese experience in a caricature. New historical studies, such as those conducted by Flynn (2006) and Stoner (2007) are providing a far more accurate perspective on the Chinese immigrants. There is much work yet to be done to understand the range of complexities associated with the history of Chinese in Montana, and it is hoped that the historical information consolidated for this book will guide future researchers. Such a "road map" did not exist before this research.

The three historical periods outlined may prove useful because the boundaries are based significantly on the social circumstances and legal situation of the Chinese in Montana, though other scholars are likely to find other ways of organizing the Chinese experience. Historical information in the 1860s and 1870s suggests that the Chinese were relatively unfettered or constrained by effective and enforced legal restrictions, though this changed with passage of the Opium Act in 1880 and the federal Chinese Exclusion Act in 1882. By 1900 placer mining made only a minor contribution to the Montana economy, and Chinese workers searched for other employment opportunities. During the mid-20th century the Chinese population was in rapid decline, down to only 10% of its size in 1870. Historical information provided not only broad patterns of Chinese heritage but also gave it a personal character, though little remains of individual experiences.

A second goal of research was to compile an exhaustive database of Chinese archaeological sites and collections in Montana. This turned out to be a far more challenging task than first thought, due to technological limitations of the State Historic Preservation Office's (SHPO) archaeological site database, known as the Cultural Resources Information System or CRIS. All archaeological forms are digitized, but there is no ability to search these forms for ethnic designations such as "Chinese." Recent programs at SHPO are starting to change this situation, though it will be many years before researchers can search the database by ethnic affiliation or other relevant keywords. The list of 40 known and 20 potential Chinese archaeological sites is the culmination of personal communications, direct searches of legal locations derived from historical data, and the partnership with Region 1 of the United States Forest Service. It is clear that many areas with high Chinese site potential were severely damaged by later remining and development. This process of site destruction continues today, as with the reclamation of the Beal Mountain Mine at the head of German Gulch. While representing a comprehensive search of public records, the sites currently within the database likely reflect only a small portion of Chinese sites, as many exist on private lands or in unsurveyed portions of public lands.

By using historical and archaeological data on the Montana Chinese experience, it was possible to analyze the nature of social organization among this immigrant population. Voluntary organizations such as clans, district organizations, and secret societies formed the most basic substitute lineage in diaspora communities and deeply affected the daily lives of Chinese immigrants in Montana. Archaeologists have yet to frame their research within the context of Overseas Chinese social organization. There is much work to be done in Montana and other diaspora communities, in order to provide a richer analysis of this topic. Historical sources about the Chinese in Montana possess only small fragments of information on voluntary organizations, including secret societies, and this information helped to interpret the population's experience in the state. The Chinese in Montana used these voluntary organizations as a major adaptive mechanism for resisting exclusion and racism in diasporic communities, and these activities deeply affect the interpretation of archaeological materials. Instead of purely a Chinese versus non-Chinese dichotomy, exploring internal social organization provides a far more nuanced view of the shaping of culture and its material manifestations. In Montana the Chinese organized themselves in voluntary organizations that helped to dictate the form of their resistance to exclusion laws, boycotts, and institutional racism. At the same time, these institutions provided the context and mechanism for elaborate trade networks that brought material goods from their home in China in mass quantities to the most isolated parts of Montana.

## *Entrance and Expansion of the Chinese in Montana (1862–1880)*

Chinese immigrants first came to Montana after the gold discovery at Bannack in 1862. By 1870 the Chinese population exploded, accounting for 10% of the territory's total population and even higher proportions in urban areas such as Virginia City and Helena. The period between 1862 and 1880 is typified by the domination of placer mining as the main economic pursuit for the Chinese immigrant population. Far from modern centers of population, Chinese dominated the mining communities in Cedar and Gold creeks, and German and Cave gulches throughout the late 1860s and 1870s. Unlike in later periods, legal challenges to the rights of Chinese immigrants were largely ineffective during the first two decades of Montana Territory. Montana's Alien Law, passed in 1872, failed to alter much, as many European American miners preferred to sell under-paying claims to Chinese companies for substantial profits. Though the Chinese population felt the pains of a nationwide depression of the 1870s, their population dropped only slightly by 1880.

Research of archaeological sites from this first historical period demonstrates the ubiquity of Chinese populations in isolated mining camps like Lion City, Cave Gulch, Ruddville, and Yogo Town. German Gulch is still the premier archaeological site in Montana due to its reflection of community organization, its isolation, and its overall high level of integrity. German Gulch best reflects the archaeological and historical conditions of the Chinese during the 19th century, with the creation of a nearly self-contained society with its own store, recreational pursuits, and well-paying gold claims. The archaeological recovery of imported Chinese goods, from coconuts to porcelain serving bowls, illustrates the extensive trade networks existing in Montana before the first railroad even entered the territory. In apparent affront to the Montana Alien Law, the Chinese owned most of the placer claims in German Gulch during the 1870s and made a substantial profit. By the end of the 19th century, however, German Gulch's placer streaks played out, and without any other industry to supplement their income, the settlement faded into Montana's history.

Other sites in this first period further illustrate an open frontier of opportunity, surrounded by the institutional racism inherent in Montana and other parts of the American West. Archaeological work in the Cedar Creek drainage of western Montana recovered evidence of Chinese lives during the 1870s at China Gulch, Louiseville, and the Rowley Chinese Hearths. Due to the anti-Chinese agitation in neighboring Idaho, hundreds of Chinese miners and entrepreneurs moved into Cedar Creek during the summer of 1870. Archaeological research has shown that many of the initial Chinese moving into China Gulch lacked the economic clout to purchase sufficient foodstuffs, and some individuals resorted to starvation processing of faunal remains. Meanwhile, by the mid-1870s, the lot of Chinese in Cedar Creek changed and the archaeological deposits in Louiseville demonstrate a broad palette of food and imported Chinese goods.

*Legal Challenges and Triumphs of the Chinese in Montana (1880–1900)*

The last two decades of the 19th century witnessed the peak and collapse of the Chinese population in Montana due to the changing legal and economic circumstances of the state. Dipping slightly from 1870 to 1880, the Chinese population of Montana peaked in 1890 thanks in large part to the construction of the Northern Pacific Railroad. H. L. Hallett, supervisor for the Clark's Fork Division of the Northern Pacific, employed upward of 6,000 Chinese railroad workers through Washington, Idaho, and Montana during the first years of the 1880s. Nowhere was this population larger than in Montana, with thousands of Chinese working the grade and laying track from the Idaho border eastward to Gold Creek. The successes of the Chinese on the Northern Pacific were overshadowed by a changing legal climate in Montana and the United States.

Montana's Opium Act, passed in 1880, targeted the Chinese use of smoking opium while protecting other forms of opium consumption generally used by European Americans. Unlike the Alien Law, the Opium Act was strictly enforced and led to even further damage to the popular opinions of Chinese immigrants. The Chinese Exclusion Act of 1882 was the first in a pattern of anti-Chinese legislation that dominated the last two decades of the 19th century. The Exclusion Act, coupled with the provisions of the Geary Act amendment in 1892, effectively ended the legal immigration of Chinese women and laboring class of men into the United States. In addition the Geary Act forced Chinese to register and led to a number of high profile deportations around the nation. Concurrent with the immigration legislation was a growing, and increasingly violent, anti-Chinese movement spearheaded by powerful labor unions. Labor union–supported boycotts, anti-Chinese immigration legislation, and changing economic circumstances signaled the end of a robust and growing Chinese population in Montana. By 1900 the Chinese population dropped and never recovered.

Archaeological research into the second period of Chinese history in Montana is not as thorough as for the other periods. While German Gulch's occupation crosses into the 1880s and early 1890s, the Chinese population of the district had declined significantly from the 1870s heyday. Archaeological research completed on sites dating from this period reflects the transition from the dominance of mining-based employment of the Chinese to a diversification into laundries, restaurants, and other service-based industries. Excavations of the Chinese laundries at Marysville demonstrate how these small isolated ethnic populations continued to connect their lives to broader patterns of history by supporting the Chinese Empire Reform Association and maintaining traditional cultural practices.

On the fringes, some Chinese miners continued to work placer claims into the late 1890s and even early 1900s, though with limited success. The discovery of the Quong Chong Company claim outside Yellowstone National Park reflects the last vestiges of Chinese activities in placer mining during the 19th century. Due to the site's isolation and the lack of Chinese markets nearby, the archaeological assemblage consists largely of European American ceramics and bottles. This assemblage, however, does not necessarily reflect the acculturation of Chinese to European American culture but rather the miners' isolation from others in their ethnic community and their ability to adapt their traditions to meet changing circumstances.

The history of the Chinese during the last two decades of the 19th century in Montana reflects a crowning achievement of their labor in the construction of the Northern Pacific Railroad. While archaeologists have not completed a detailed survey of the grade, the work of avocational archaeologist Gary Weisz located eight distinct railroad camps associated with the construction. These sites, as part of a work camp system, imprint a highly regular archaeological signature from Idaho to Missoula. Weisz found the same suite of Chinese ceramics, opium cans, and opium paraphernalia at all the recorded camps, further illustrating these cookie-cutter camps that moved along the grade during the early 1880s. More interesting, perhaps, is the continued dominance of Bamboo-style ceramics well after their peak in the early 1870s in other parts of the American West. The material culture at these camps reflects the creation of a cultural system from a distributor tied to a certain range of goods that had to meet the demands of the workers without catering to their every whim or desire.

Chinese voluntary organizations played a significant role in shaping the experience of this immigrant group during the late 19th century. The Six Companies, with direct ties to secret societies, and other Chinese voluntary organizations helped the Northern Pacific Railroad to organize their labor needs. In addition several federal laws excluding the Chinese in Montana and other parts of the United States were resisted explicitly by voluntary organizations led by the Six Companies based out of San

Francisco. Voluntary organizations provided at least some level of political organization designed to fight back against these racist laws, and there was a dedicated effort to resist the Geary Act amendment in 1892 through a registration boycott. Chinese voluntary organizations affected the ways in which the Chinese acquired goods and services, as revealed in the remarkable similarity of archaeological materials across space and time during this period. Many archaeologists viewed the continued use of Chinese goods as a sign of social isolation and perhaps resistance to assimilation. In reality the social system affected the types of goods available to purchase through Chinese vendors, reinforced by the social implications of refusing to acquire supplies through the leader (generally a merchant) of the local or even regional voluntary organization.

### *Aging, Diversification, and Population Collapse of the Chinese in Montana (1900–1943)*

While the peak of Chinese population in Montana had passed, the first decades of the 20th century continued to include substantial contributions by this maligned immigrant population. During the 1880s and 1890s the Chinese continued to shift away from mining as the dominant means of economic gain. By the 20th century the Chinese in Montana had created a diversified economic model centered on service-based industries like laundries, restaurants, and mercantile shops. Service industries required the least amount of capital investment and provided means to accumulate money in the new economy of the 20th century. Due to the exclusion of Chinese from hard-rock mining and other industries by labor unions, there were fewer opportunities in the extractive part of the economy. Chinese also worked as gardeners, doctors, tailors, stock raisers, and in numerous other jobs that break with the common stereotypes of Chinese as laundry or restaurant workers. Even with all the past racist transgressions, the Chinese in Montana supported the United States effort in World War I, even establishing a War Bonds Society in Big Timber. As the 20th century progressed, the Chinese population of Montana dwindled and aged, with an ever-growing population of Japanese immigrants in its place (Figure 26). The few hundred remaining Chinese witnessed the passing of their community in Montana, with many returning home to China or dying. Some of the Chinese left in Montana had immigrated to the United States in the 1870s and had not seen their wives, children, or other family members for decades. Effectively, Montana became home and a final resting place for hundreds

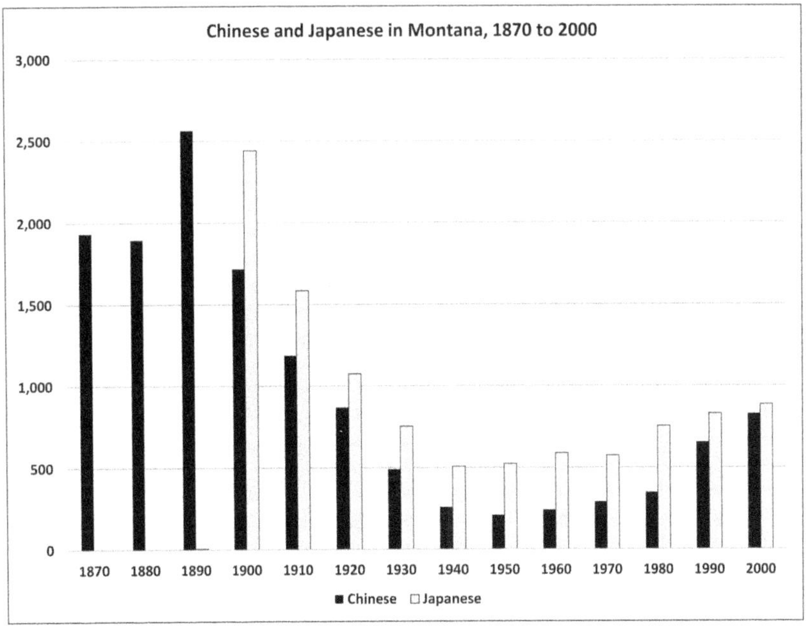


Figure 26: Chinese and Japanese population of Montana, 1868–2000. (Created by author.)

of Chinese immigrants. There are still sections in many Montana cemeteries that silently testify to the life and death of this immigrant population.

While Montana's Chinese population declined throughout the first five decades of the 20th century, isolated businesses continued to prosper. Archaeological efforts in Butte and Big Timber recovered significant information on these businesses and social relations. University of Montana archaeologists discovered the remains of a Chinese laundry and restaurant in Big Timber, and the information recovered from that site support Rose Hum Lee's (1949) outline of Chinese business models. A cursory analysis of the faunal remains, ceramics, and bottle assemblage does suggest that this particular Chinese establishment catered to a diverse clientele, largely European American. The presence of Chinese ceramics and opium paraphernalia in the back of the structure suggests a public versus private dichotomy of activity at the business. In addition the periodic switching of the business from laundry to restaurant illustrates a remarkable flexibility in adapting to new economic circumstances.

Excavations at Butte discovered several large trash middens associated with Chinese occupation and reflected households with diverse socioeconomic backgrounds, all located within a single city block. The single block in Butte contained several Chinese businesses, habitations, and even the Baptist Mission that promoted the education and assimilation of the immigrant population. Due to the population decline the Chinese clustered into these dense occupations, where individuals or households of diverse socioeconomic backgrounds lived in close proximity. The wide variety of goods discovered in Butte likely results from over 60 years of established Chinese trade networks and the diversity of the area's occupants.

Finally, the archaeological remains at Forestvale, Mount Moriah, and Philipsburg cemeteries provide closure to the Montana story for those Chinese unable to return home. At Forestvale Cemetery in Helena, more than 200 Chinese were interred, and mourners constructed a funerary burner that was employed for over 50 years. Mount Moriah Cemetery in Butte and Philipsburg City Cemetery both have remnants of funerary burners, providing additional remains reflecting the continuation and creolization of Chinese mortuary rituals. For instance, Chinese mourners sacrificed both traditional goods and European American goods at the funerary burners, reflecting a symbolism not tied to specific presentation or origin of artifacts. Many of the Chinese who pioneered Montana remain entombed in its soil.

The unique facets of Chinese social organization played out in the last period of Overseas Chinese history in Montana to 1944, affecting the personal lives of many immigrants. As in previous decades, these voluntary organizations provided funds for the funerals, burials, and even transmission of the bodies back to China. As seen in the case of Wong Chuck's murder of Mow Lu Dun in Helena in the 1920s, Chinese individuals in Montana had deep and sometimes nefarious ties to secret societies along the Pacific Coast and in mainland China. For a Chinese merchant to arrive from Seattle and pay for the burial of Mow Lu Dun, and then push for the release of Wong Chuck, suggests efforts to maintain the fragile peace between clans, district associations, or secret societies. Due to the decrease in Chinese population during the 20th century, the overall effects of social organization were not as pronounced as in earlier periods. It is clear that voluntary organizations helped to foster social cohesion within the Chinese community in Montana, and sometimes competing voluntary organization interests spilled over into violent episodes fought along membership lines. Butte had a number of these disputes in the early 1900s, which resulted in several deaths (Lee 1978).

## *Chinese Overseas Social Organization in Montana*

While many may see the Montana Chinese story as typified by racism, legal exclusion, and violence, their success in overcoming these circumstances is the true narrative, as it was elsewhere among 19th- and 20th-century diaspora communities. As noted by Lyman (1974), Overseas Chinese populations were connected by a complex web of voluntary social organizations created through clans, district organizations (or benevolent societies), and secret societies. All having roots in China, the new forms of voluntary organizations in diaspora communities provided the immigrant with a substitute for the traditional family-based social organization. These organizations served a variety of formats, and the lines between them were often blurry. These organizations deeply affected the lives of Chinese immigrants in Montana and beyond. Specifically, secret societies, illegally formed in China to subvert the Qing Dynasty, found an unfettered environment for growth in influence and wealth in Overseas Chinese communities. Merchants, traditionally excluded from power in pre-1911 Chinese society, quickly rose to dominate all three types of voluntary organizations. Most significant for archaeologists, these merchants controlled the flow of people and goods throughout the American West and other regions affected by the Chinese diaspora. These voluntary organizations provided the Chinese immigrant with a means of protection and power and likely allowed the degree of success witnessed in Montana and beyond. What scholars have called an insular Chinese society in the United States was the result of internal social organizational structures and their connections to external racist and legal factors.

The Chinese in Montana employed an interrelated web of social relations that replaced the traditional kinship systems in China (Figure 27). Clans, district associations, and secret societies helped to allow the Chinese to succeed in Montana by organizing communal mining efforts, presenting resistance to exclusionary laws and boycotts, and even controlling the flow of people and goods throughout the state (Figure 28). It appears that Helena, Virginia City, and Butte were the centers of Chinese secret society activity throughout the 19th century. By the 20th century the Chinese Empire Reform Association spread out to most of Montana's Chinese communities. Voluntary associations did far more than merely allow survival in the United States or Montana; they were powerful agents of change within the Chinese community overseas and integral to reformation and revolution in China itself.

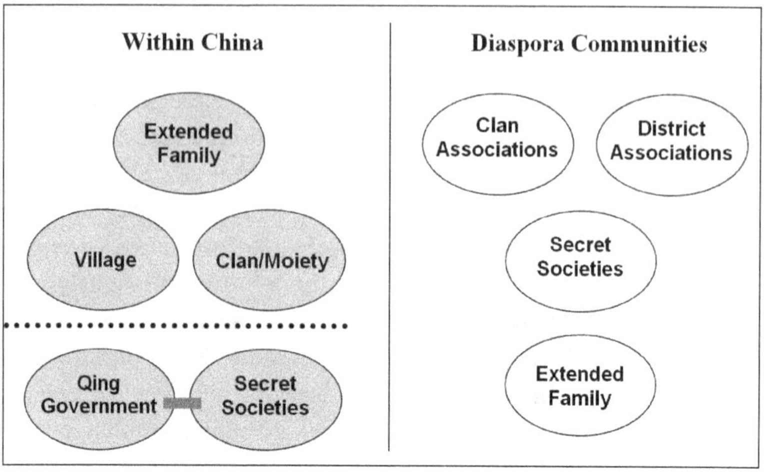


Figure 27: Traditional social organization in China (*left*), and new social organization opportunities in Chinese diaspora communities (*right*). In China the extended family, village, and clan represented the most significant form of social organization, with the Chinese government existing at a higher level. In China secret societies were at odds with the Qing government, hence the connecting link. Chinese immigrants in diaspora communities continued the linkage of extended families, but due to restricted immigration there was less ability to retain those ties. Thus voluntary associations (clans, district, and secret societies) were the most prominent means of social organization. (Created by author.)

As archaeologists work primarily with the material culture of past peoples, it is important to understand that the nature of Chinese social organizations deeply affected artifact patterning by affecting the availability of goods in a given context. For instance, as wealthy Chinese merchants headed most of the voluntary organizations, these individuals dictated the flow of goods into a mining camp or even urban center. Future research can perhaps track down the linkages of trade between Seattle, San Francisco, China, and points in Montana to determine any networks built upon these voluntary organizations and merchants. While experiments have yet to be carried out, chemical source analysis of Chinese porcelain, opium pipe bowls, medicine bottles, imported flora and fauna, and other artifacts could indicate further linkages of trade networks based upon district, family, or secret society ties. While Chinese ceramics share the same morphological characteristics, they were produced in various areas of China and were transmitted by differing trade networks that could reflect voluntary organization influences or economic ties.

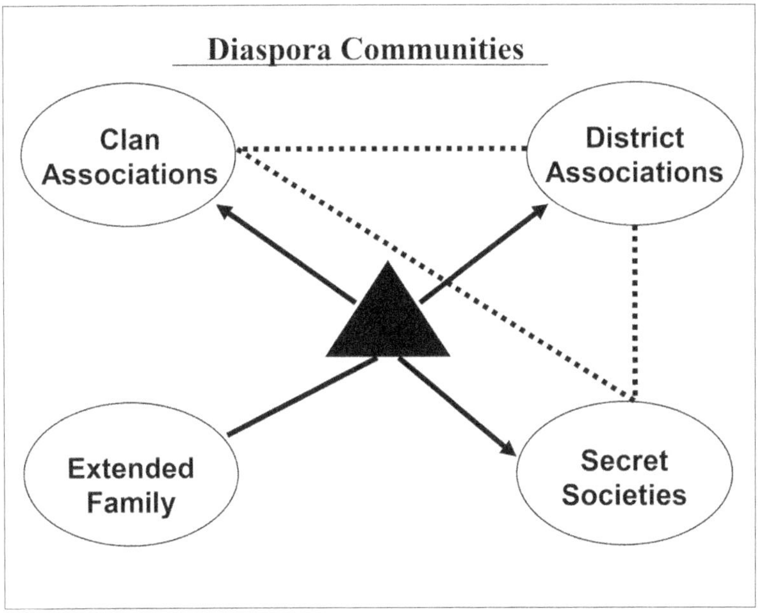


Figure 28: Social organization opportunities available to the Chinese immigrant in diaspora communities. The individual (indicated by the triangle in the center) had direct linkages to the extended family. This same individual could choose which voluntary organization(s) to join (hence the uni-directional arrows). Clan associations were extensions of similar organizations in China and were centered on a single lineage or surname group. District associations cast a broader net but were based on ethnicity, county, and sometimes language groupings. Secret society membership, however, was based purely on willingness to belong to the group and cross-cut the more traditional roles of family, lineage, clans, language, and ethnicity. All Chinese voluntary organizations had overt and hidden linkages of joint membership or control and, in some cases, animosity (denoted by the dashed lines). Thus an individual who chose to belong to a single group could be indirectly linked to the other two types of voluntary organizations. During conflicts this complex web of relations caused most of the Chinese population in a certain area like San Francisco to be drawn into a conflict. (Created by author.)

## Myths of the Montana Chinese

It is clear from the historical and archaeological analysis of the Chinese in Montana that this population was as diverse and internally heterogeneous as any other immigrant group. Certain myths persist about the Chinese experience, and as most of the descendant population no longer lives within the state, these rumors can grow rampant without much substantiation. Another outgrowth of the research, which failed to fit neatly into the narrative, is the role myth plays in our current understanding of the Chinese experience in Montana. Many modern Montanans' only experience with the Chinese is through stories that generally perpetuate stereotypical images of the Chinese as clever, thrifty, hard-working, and amusing anecdotal characters. These images play themselves out even today with many citizens discussing Chinese tunnels, rock work, and "opium bottles." A basic misunderstanding of the Chinese experience in Montana pervades mainstream education and awareness, though the Mai Wah Museum, Montana Historical Society, and Mike Mansfield Center have released educational tools to assist in dispelling some of the misinformation. The desire to preserve and interpret the Chinese heritage of the Montana even led Charles Bovey to reconstruct examples of a joss house and other Chinese businesses at Nevada City in the 1970s, using salvaged wood and logs from other historic structures (*Montana Standard* 1973:22). In reality, though, the image of Chineseness at Nevada City is a mere construction of basic stereotypical images by a prolific collector, situated in a community where little or no Chinese presence ever occurred.

The true depth of Chinese contributions to Montana remains misunderstood, with many using this population as a tourist attraction even where no heritage exists. Take, for example, the Wu Tang Chinese Drug and Laundry in Neihart that is proudly displayed in *Roadside Geology of Montana* by University of Montana professors David Alt and Donald W. Hyndman (1986:319). The possibility of such an historic log structure relating to the Chinese would be a windfall for researchers, especially considering the ordinance prohibiting Chinese from living in Neihart passed by a miners' council in 1885. During a revisit in 2008, the researcher discovered that the building burnt down in the early 2000s, and a local artist painted the "Wu Tang" text on a random miner's cabin in the 1970s to bring in tourists.

### The "Heathen Chinee"

As illustrated best by Kelly Fong (2005, 2007), the pervasiveness of Chinese stereotypes in United States history, archaeology, and collective memory continues to undermine the narrative of these pioneers. In particular, individuals discussing the lives of Chinese in diaspora communities tend to focus on their "otherness" in terms of opium use, diet, religion, and burial ceremonies.

Wylie and Fike (1993:298–199) illustrate that the Chinese use of opium differed greatly from that of host communities. Chinese used opium for a variety of purposes rather along the lines of medical marijuana in today's culture, while European Americans tended to use the drug more like alcohol. American doctors and culture viewed medicinal opium, usually in a liquid form mixed with alcohol, as an acceptable and necessary cure for physical ailments, though it possessed the same ability to addict its user (Ahmad 2007:38). Indeed, Chinese were not the only people smoking opium in the United States and Montana, as many European Americans, African Americans, and members of other ethnic groups became addicted to the drug during the 19th and early 20th century (Ahmad 2007:30, 90). American cultures viewed smoking opium as far more pernicious, with users losing their ability to employ morals or religion, and they connected the drug with a feminine trait of docility, whereas alcohol was regarded as far more masculine and thus healthy (Ahmad 2007:38)

Fong (2005:23) notes how archaeologists continue the trend of focusing on the cuisine preferred by the Chinese, and these discussions continue to place this population at odds with mainstream American culture. In fact the Chinese practiced a synergy of traditional cuisine and American foods dependent upon availability of goods. Many Chinese viewed the consumption of beef with the same wariness as European Americans felt upon the consumption of dogs and cats. Every region in the United States has specific foods that promote local identity, and the Chinese were no different in this.

The most dominant narrative found in historical resources is the focus on using the epithet "heathen" to describe the Chinese population in the United States. Most Chinese immigrants came to the United States with a firm faith in a unique mixture of Daoism, Buddhism, and Confucianist belief systems (Rouse 2005:21–22). In many Montana communities the Chinese established temples for worship while their European American counterparts were still building saloons. Beyond this, some of the Chinese who immigrated to the United States in the 19th and 20th centuries were actually Christians who had been missionized by Protestants in China.

Many Chinese who died in Montana during the 19th and 20th centuries performed traditional burial rites at public or private cemeteries, much to the amusement of newspaper reporters and local citizens. As best illustrated by Rouse (2005:40–41), many Americans criticized the Chinese ritual offerings of food and drink for the deceased, while at the same time leaving flowers at their own relatives' graves. During some Chinese burials in Montana and elsewhere (Rouse 2005:35), many of the European American funeral attendees fought over the food, liquor, and money left behind by the mourners. Another persistent myth regarding the Chinese is that all deceased have been exhumed and shipped back to China for reburial. This rationale has led to the abuse and neglect of traditional Chinese burial sites in Montana, even though there are likely of dozens of bodies remaining. Historical and archaeological evidence does indicate that bodies were disinterred and returned to China, but there were many more bodies of criminals, paupers, and prostitutes that were generally not afforded this luxury (Chung and Wegars 2005:13).

### *Chinese Rock Work*

There continues a persistent belief in the United States, Canada, New Zealand, and Australia that nicely stacked placer tailings are an indicator of Chinese miners. Following a position similar to that of Lawrence (2000), McGowan (2005) asserts that neatly stacked rock from placer mining relates specifically to Chinese miners. McGowan (2005) argues against Valentine's (2002:47–49) statement that well-sorted mine tailings are ethnically indistinct. According to McGowan (2005:17), "vertically stacked tailing mounds, in particular, are very strong evidence of Chinese mining activities." This goes against scholarly convention in the United States, as Valentine (2002) notes that neatly stacked rock cannot necessarily be attributed to a specific ethnic group, even if oral history and legends relate extensive rock work to the Chinese. The fallacy of Chinese being the only individuals capable or industrious enough to stack rock tailings neatly has cost many researchers in the United States time and money (e.g., Merritt and McLeod 2010). The Poacher Gulch Terraces project, which was the initial spur of this research, was thought to be Chinese due to the hand-stacked dry laid rock work, though no evidence could ever be linked to this hypothesis. Fredlund and colleagues (1991) studied the mining works in German Gulch, Montana, and while they could clearly see differences in sluice-box placering remains versus hydraulic mining tailings, there was no clear ethnic differentiation. The mere idea that the Chinese stacked rock better than European Americans on the mining frontier is steeped in racial stereotyping and still affects the public perception of the Chinese today.

### Chinese Tunnels

Another feature commonly associated with the Chinese both in Montana and elsewhere in the United States and Canada is presence of subterranean tunnels (see Asian American Comparative Collection 2007). This is perhaps the most persistent local urban legend in Montana in regard to the Chinese; and during the course of this project tunnels have been described to me in Missoula, Livingston, Big Timber, Havre, Virginia City, Butte, Hamilton, Kalispell, Billings, and Helena. As common as the story of the tunnels themselves is the constant refrain that the tunnels have been destroyed and thus cannot be relocated. During the archaeological project at Big Timber in 2008 researchers investigated potential "Chinese tunnels" without any success, even with the assistance of a backhoe (Moschelle 2009). It appears that most of the purported tunnels in Montana's communities are sidewalk vaults, steam tunnels, or underground store fronts. In many historic communities there are voids under sidewalks for use as underground storage for the businesses above and, in some rarer cases, even underground store fronts during a community's early urbanization. Steam tunnels were common in the American West to convey heat and steam from boiler plants to businesses throughout a community's downtown, and were later used as a void for both legal and illegal multiple purposes, but not solely for the Chinese. Lai (1991:39) provides a solid overview of how the myth of Chinese tunnels grew in Victoria, British Columbia, and this same pattern applies to Montana.

This is not to say that the Chinese did not use any subterranean spaces in Montana communities. Instead, it appears that the Chinese and many other groups adaptively used underground spaces at various times for diverse reasons. But to assume that the Chinese constructed tunnels for living and business is folly and is an extension of a racially stereotyped image of these immigrants. A blatant example of racist perspectives comes from an underground tourist attraction in northern Montana, where the dominant narrative centers on the Chinese being forced to live underground by the Great Northern Railroad Company for their own protection from marauding European American mobs. This story is suspect on a number of counts, including that the Japanese were the main non–European American labor force employed on the Great Northern and that the period discussed is decades after the peak of anti-Chinese violence in the United States. The myth of the Chinese tunnels will be a powerful and persistent myth for professionals to wrestle with, as a resurgence in interest has occurred lately (Vollmann 2004)

### *Chinese Hearths and "Ovens"*

"Chinese ovens" are a common discovery by locals in Montana communities near railroad grades. This topic has been well covered by Priscilla Wegars (1991) and need not be discussed in detail. Many historical societies in Montana already possess a copy of Wegars's article, and her efforts have helped educate numerous interested Montanans. In most cases the common so-called Chinese ovens described by locals were built by Eastern Europeans or Italians during the construction phases of Montana's Northern Pacific, Great Northern, and Milwaukee railroads. Locals discovered an "oven" east of Paradise, Montana, along the Northern Pacific Railroad grade, and it is an excellent example of the types of features commonly misinterpreted as Chinese in origin. As shown in detail at China Gulch in western Montana, it appears that most Chinese hearths are U-shaped with an open front for access needed in using traditional wok cooking techniques.

## ***Hua Kiu***

The lives of Overseas Chinese, or *Hua Kiu*, are deeply intertwined with the historical development of Montana, though their contributions remain forgotten among all but a few dedicated scholars and a small number of Montanans. So much of the Chinese story remains untold due to their lack of a voice in Montana's history. Chinese migrants to Montana spurred economic growth through mining and the construction of the Northern Pacific Railroad, and their population grew with the state's rural and urban development. Facing racism through legal restrictions, violence, and boycotts, the Chinese of Montana nevertheless prospered, some returning to China wealthy. Other Chinese individuals fared less well, became part of the fabric of local history, and remain buried in Montana today, not infrequently in overgrown and neglected cemeteries. With every trowel scrape and newspaper extract, the story of the Montana Chinese unfolds, but only through brief and glancing views. These archaeological and historical glimpses provide snapshots of regional and international patterns of an immigrant group's adaptation to changing social and economic circumstances. Using both modified and new forms of social organization,

Figure 29: "Chinatown" road sign on Jeff Davis Creek southwest of Dillon, Montana. (Photo by author, 2008.)

the Chinese in the United States and Montana were collectively able to resist the seemingly insurmountable racism and legal exclusion facing them. While most of the population left Montana in the early 1900s, their legacy continues in the historical and archaeological remnants of their experiences. The Chinese were not anecdotal appendices to the settlement and exploitation of Montana but, instead, were a part of the master narrative largely ignored until only recently. To remember is to know, to know is to respect, and respect is what these coming men and women from China deserve. The Chinese never truly left Montana, as their influences and achievements still surround modern Montanans. Place names referencing people and things long since forgotten provide reminders of Montana's historic Chinese pioneers. So much more research needs to be completed into the Chinese experience, and this book represents only one small part of a larger need.

# References


Ahmad, Diana L.

2007 *The Opium Debate and Chinese Exclusion Laws in the Nineteenth-Century American West.* University of Nevada Press, Reno.

Akin, Marjorie K.

1992 The Noncurrency Functions of Chinese Wen in America. *Historical Archaeology* 26(2):58–65.

Allen, Frederick

2004 *A Decent Orderly Lynching: The Montana Vigilantes.* University of Oklahoma Press, Norman.

Allen, Rebecca, R. Scott Baxter, Anmarie Medin, Julia Costello, and Connie Young Yu

2002 *Excavation of the Woolen Mills Chinatown, San Jose*, 2 vols. Prepared for California Department of Transportation, Oakland.

Alt, David, and Donald W. Hyndman

1986 *Roadside Geology of Montana.* Mountain Press Publishing Company, Missoula MT.

*Anaconda Standard*

1900a Funeral of Thomas Gogk. May 4, p. 12.

1900b Obituary Notice. September 2, p. 12.

1900c Letters of Administration, Missoula. September 17, p. 10.

1900d His Foot Was Frozen, Missoula. November 20, p. 12.

1901a Gypsies Reach Missoula. June 21, p. 12.

1901b Raided the Chinaman. August 16, p. 12.

1901c Four Suspects Arrested. August 19, p. 10.

1901d Chinamen Robbed Again. October 1, p. 12.

1907 George Young, Chinaman, Will Be Buried Today. May 31, p. 7.

Arkush, R. David, and Leo O. Lee (eds.)

1993 *Land Without Ghosts: Chinese Impressions of America from the mid-Nineteenth Century to Present.* University of California Press, Berkeley.

Asian American Comparative Collection Website (AACC)

2007 Chinese Tunnels. University of Idaho, Moscow, http://www.uiweb.uidaho.edu/ls/aacc/research.htm.

Avon Get Together Club

1977 *Our Neighborhood.* Avon MT.

Baxter, Jane Eva

2005 *The Archaeology of Childhood: Children, Gender, and Material Culture.* Rowman AltaMira Press, Walnut Creek CA.

Baxter, R. Scott

2008 The Response of California's Chinese Populations to the Anti-Chinese Movement. *Historical Archaeology* 42(3):29–36.

Beaverhead County History Book Association (BCHBA)

1990 *The History of Beaverhead County*, 2 vols. Beaverhead County History Book Association, Dillon MT.

Benn, Charles D.

2002 *Daily Life in Traditional China: The Tang Dynasty*. Greenwood Publishing Group, Westport CT.

Bennitt, Mark

1905 *History of the Louisiana Purchase Exposition*. St. Louis Louisiana Purchase Exposition Company, St. Louis MO.

*Benton Record*

1880 Extreme Vigilance. February 13, p. 2.

1881a Raid on an Opium Den. June 16, p. 2.

1881b Collector T. A. Cummings. June 16, p. 2.

1881c All the Chinamen, but One. June 16, p. 2.

1883 Chinamen Seem to Be the Best. April 28, p. 2.

Bernstein, David E.

1999 Lochner, Parity and Chinese Laundry. *William & Mary Law Review* 41(1).

Bik, Patricia

1993 China Row: A Report on the Chinese Burial Ground at Forestvale Cemetery. Prepared for the Forestvale Cemetery Board of Lewis and Clark County. On file, Montana Historical Society Archives, Helena.

Binford, Lewis R.

1962 Archaeology as Anthropology. *American Antiquity* 28(2):217–225.

Blackford, Lindsay, and Christopher W. Merritt

2008 *Granite Chinatown Artifact Inventory*. On file: Montana State Historic Preservation Office, Helena.

Blake, Henry N., and Cornelius Hedges

1877 *Reports of Cases Argued and Determined in the Supreme Court of Montana Territory*, Vol. 2. Sumner Whitney and Company, San Francisco CA.

Blanford, John

1987 Overseas Chinese Ethnicity and European American Glass Bottles at the Wong Ho Luen Site. In *Wong Hu Leun: An American Chinatown*, 2 vols., pp. 2:189–232. Great Basin Foundation, San Diego CA.

Bloch, Maurice

1983 *Marxism and Anthropology*. Oxford University Press, Oxford.

Boas, Franz

1897 *The Social Organization and the Secret Societies of the Kwakiutl Indians*. United States National Museum, Government Printing Office, Washington DC.

Bockhorst, Andrew

2003 *The Sino-Sonoran: Site and Artifact Variability of the Chinese in Southern Arizona*. Master's of Arts thesis, Department of Anthropology, University of Arizona, Tucson.

Bolz, Jennifer

1995 Chinese Organized Crime and Illegal Alien Trafficking: Humans as a Commodity. *Asian Affairs* 22(3):147–158.

Boswell, Terry E.

1986 A Split Labor Market Analysis of Discrimination against Chinese Immigrants, 1850–1882. *American Sociological Review* 51(3):352–371.

Bowen, Kristin

2004 *The Chinese Presence in Virginia City, Montana: A Historical Archaeology Perspective.* Master's thesis, University of Montana, Missoula.

Brettell, Caroline B.

2005 Voluntary Organizations, Social Capital, and the Social Incorporation of Asian Indian Immigrants in the Dallas–Fort Worth Metroplex. *Anthropological Quarterly* 78(4):853–883.

Brook, Timothy

1990 Family Continuity and Cultural Hegemony: The Gentry of Ningbo, 1368–1911. In *Chinese Local Elites and Patterns of Dominance*, J. Esherick and M. Rankin, eds., pp. 27–50. University of California Press, Berkeley.

Burlingame, Merrill G.

1942 *The Montana Frontier.* Montana Historical Society, Helena.

*Butte Miner*

1922 Tong War. February 17, p. 3.

Chan, Sucheng (ed.)

2006 *The Flow of People, Resources, and Ideas Between China and America during the Exclusion Era.* Temple University Press, Philadelphia PA.

Chen, Shehong

2006 Republicanism, Confucianism, Christianity, and Capitalism in American Chinese Ideology. *The Flow of People, Resources, and Ideas Between China and America during the Exclusion Era*, Sucheng Chan, ed., pp. 174–193. Temple University Press, Philadelphia PA.

Chen, Ying-ying

2001 *In the Colonies of Tang: Historical Archaeology of Chinese Communities in the Northern Cariboo District, British Columbia (1860s–1940s).* Doctoral dissertation, Department of Archaeology, Simon Fraser University, Burnaby, British Columbia.

Chesneaux, Jean

1972 Secret Societies in China's Historical Evolution. In *Popular Movements and Secret Societies in China 1840–1950*, J. Chesneaux, ed., pp. 1–22. Stanford University Press, Stanford CA.

Cheung, Floyd

2002 Performing Exclusion and Resistance: Anti-Chinese League and Chee Kung Tong Parades in Territorial Arizona. *Drama Review* 46(1):39–59.

Chiu, Ping

1967 *Chinese Labor in California, 1850–1880.* State Historical Society of Wisconsin, University of Wisconsin, Madison.

Ch'u, Tung-Tsu

1962 *Local Government in China under the Ch'ing.* Stanford University Press, Stanford CA.

Chung, Sue Fawn

1998 Their Changing World: Chinese Women on the Comstock, 1860–1910. In *Comstock Women: The Making of a Mining Community*, Ronald M. James and C. Elizabeth Raymond, eds., pp. 203–228. University of Nevada Press, Reno.

Chung, Sue Fawn, and Priscilla Wegars (eds.)

2005 *Chinese American Death Rituals: Respecting the Ancestors.* AltaMira Press, Lanham MD.

Ciabattari, Andrea

1976 Memories Linger, Long after Chinatown's Gone. *Montana Standard*, July 4, p. 16.

*Colored Citizen*

1894 Having seen. October 1, p. 1.

Crane, Paul, and Larson Alfred

1940 The Chinese Massacre. *Annals of Wyoming* 12(1):47–55.

Culin, Stewart

1890 Chinese Secret Societies in the United States. *Journal of American Folklore* 3(1):39–43.

*Daily Missoulian*

1900 Missoula's Opium Den. May 20, p. 3.

*Daily Rocky Mountain Gazette*

1870 Chinese Newspaper. September 18, p. 2.

Daniels, Roger (ed.)

1979 *Anti-Chinese Violence in North America: An Original Anthology*. Arno Press, New York NY.

Daniels, Rudolph L.

2000 *Trains Across the Continent: North American Railroad History*. Indiana University Press, Bloomington.

Dary, David

2008 *Frontier Medicine: From the Atlantic to the Pacific, 1492–1941*. Random House, New York NY.

Davis, Deborah J.

1982 *Anti-Chinese Sentiment Found in the Missoula and Cedar Creek Pioneer: Sept. 15, 1870 to Feb. 8, 1873*. Senior seminar paper, School of Journalism, University of Montana, Missoula.

Davis, Fei Ling

1971 *Primitive Revolutionaries of China: A Study of Secret Societies in the Late Nineteenth Century*. University of Hawaii Press, Honolulu.

Davis, Jean

1966 *Shallow Diggin's: Tales from Montana's Ghost Towns*. Caxton Printers, Caldwell ID.

Davis, Wiley S.

1967 *The Chinese in Virginia City, Montana 1863–1898*. Unpublished transcription of remarks to the Montana Institute of the Arts History Group. On file, Montana Historical Society, Helena, and Madison County Historical Society, Virginia City MT.

Dean, Patty, ed.

2009 No. 30: Chinese Gardens. *Drumlummon Views* 3(1):347–352.

Deng, Gang

1999 *The Premodern Chinese Economy: Structural Equilibrium and Capitalist Sterility*. Routledge, London, UK.

*Deseret Evening News*

1900 A Federal Court Decision. March 12, p. 4.

1904 Round-Up of Chinese. July 20, p. 4.

Didriksen, Evelyn, and Bob Didriksen

1994 *Gold Pans and Singletrees*. Anderson Publications, Fairfield MT.

Diehl, Michael, Jennifer A. Waters, and J. Homer Thiel

1998 Acculturation and the Composition of the Diet of Tucson's Overseas Chinese Gardeners at the Turn of the Century. *Historical Archaeology* 32(4):19–33.

*Dillon Tribune*

1883 Some Years Ago. January 5, p. 1.

1913 Tom Haw Has Passed Away. August 1, p. 2.

Dimsdale, Thomas

1977 *The Vigilantes of Montana*. University of Oklahoma Press, Norman.

Dirlick, Arif, and Malcolm Yeung (eds.)

2001 *Chinese on the American Frontier*. Rowman and Littlefield, Lanham, MD.

Douglas, John E., and Linda J. Brown

2002 Virginia City 2000–2001: Archaeological Assessment, Volume 2: Excavation Results. Department of Anthropology, University of Montana, Missoula. Prepared for Montana Heritage Commission, Helena.

Draszt, Joan M.

1998 *Echo of Crosscuts and Rails*. J. M. Draszt, Thompson Falls, MT.

Dunbar, Seymour, and Paul C. Phillips

1927 *The Journals and Letters of Major John Owen, Pioneer of the Northwest 1850–1871*, 2 vols. Southworth Press, Portland ME.

Durkheim, Emile

1995 [1912] *Elementary Forms of Religious Life*. Reprint, Simon and Schuster, New York NY.

Eberhard, Wolfram.

1988 *A Dictionary of Chinese Symbols: Hidden Symbols in Chinese Life and Thought*. Routledge, London.

Ellis, Meredith, Christopher W. Merritt, Shannon Novak, and Kelly J. Dixon

2011 The Signature of Starvation: A Comparison of Bone Processing at a Chinese Encampment in Montana and the Donner Party Camp in California. *Historical Archaeology* 45(2):97–112.

Elvin, Mark

1973 *The Pattern of the Chinese Past*. Stanford University Press, Stanford CA.

Erlen, Jonathan, and Joseph F. Spillane

2004 *Federal Drug Control: The Evolution of Policy and Practice*. Routledge, London.

Evans, L. T.

1998 *Feeding the Ten Billion: Plants and Population Growth*. Cambridge University Press, Cambridge, UK.

Farris, Glenn J.

1979 "Cash" as Currency: Coins and Tokens from Yreka Chinatown. *Historical Archaeology* 13:48–52.

Federal Census Bureau

1870 9th Federal Census, Montana Territory. Federal Census Bureau, Government Printing Office, Washington DC.

1880 10th Federal Census, Montana Territory. Federal Census Bureau, Government Printing Office, Washington DC.

1892 11th Federal Census Compendium, Part III. Federal Census Bureau, Government Printing Office, Washington DC.

1896 Abstract of the 11th Census, 1890. Federal Census Bureau, Government Printing Office, Washington DC.

1900 12th Federal Census, Montana. Federal Census Bureau, Government Printing Office, Washington DC.

1910 13th Federal Census, Montana. Federal Census Bureau, Government Printing Office, Washington DC.

1920 14th Federal Census, Montana. Federal Census Bureau, Government Printing Office, Washington DC.

1930 15th Federal Census, Montana. Federal Census Bureau, Government Printing Office, Washington DC.

1940 16th Federal Census, Montana. Federal Census Bureau, Government Printing Office, Washington DC.

Fee, Jeff

1991 *A Dragon in the Eagle's Land: Chinese in an Idaho Wilderness, Warren Mining District, 1870–1900.* Master's thesis, University of Idaho, Moscow.

1993 Idaho's Chinese Mountain Gardens. In *Hidden Heritage: Historical Archaeology of the Overseas Chinese,* Priscilla Wegars, ed., pp. 65–96. Baywood Publishing Company, Amityville NY.

Felton, David L., Frank Lortie, and Peter D. Schulz

1984 *The Chinese Laundry on Second Street: Papers on Archaeology at the Woodland Opera House Site.* California Archaeological Reports no. 24. California Department of Parks and Recreation, Sacramento.

Firth, Raymond

1951 *Elements of Social Organization.* Watts Publishing, London.

Flaherty, Stacy A.

1987 Boycott in Butte: Organized Labor and the Chinese Community, 1896–1897. *Montana: The Magazine of Western History* 37(1):34–47.

Flannery, Kent V.

1967 Culture History vs. Cultural Process: A Debate in American Archaeology. *Scientific American* 217(2):119–122.

Flynn, Kelly

2006 *Goldpans, Guns and Grit: Diamond City from Territorial Gold Rush to Montana Ghost Town.* Hidden Hollow Hideaway Cattle and Guest Ranch, Townsend MT.

Fong, Kelly

2005 Archaeology and the "Rat-eating Chinee": The Role of Stereotype in Chinese American Site Analyses. Senior honors thesis, Department of Anthropology, University of California, Berkeley.

2007 Return of the "Heathen Chinee": Stereotypes in Chinese American Archaeology. In *Chinese America: History and Perspectives,* pp. 115–118. Chinese Historical Society of America, San Francisco CA.

*Fort Benton River Press*

1881a Sassanach Says (and Who Is Sassananch?). January 26, p. 2.

1881b Sassanach! January 26, p. 2.

1883 In Our Fruitless Search. October 16, p. 2.

1890 Opium Den Raided. September 10, p. 2.

1894a Chinese Register. February 21, p. 2.

1894b Fort Benton Photographer. February 28, p. 2.

1901a Lee Sam Deported. May 22, p. 1.

1901b Mar Joe Deported. October 16, p. 1.

1902 Ung Toy or Louis Kim. January 22, p. 1.

1904 Lee Sing in Custody. March 2, p. 1.

Fort Leavenworth Prisoner Files

1926 Yee Sing, Inmate #23147. National Archives, Kansas City MO.

Foster, Lance

1992 *The Ophir Creek Historic Mining District: A Contextual Study*. Publication no. RI–93–19. United States Forest Service, Helena National Forest, Helena MT.

Fredlund, Lynn, Dale Decco, Gene Munson, and Jean Hourston-Wright

1991 *Archeological Investigations in the German Gulch Historic District (24SB212): A Historic Chinese and Euroamerican Placer Mining Area in Southwestern Montana.*, 2 vols. Report prepared for Beal Mountain Mining, GCM Services, Butte MT.

Fullerton, Neil

1961 Chinese Graves. Unpublished research notes, September 6, 1961, MC76. On file, Montana Historical Society Archives, Helena.

Gao, Weinong

2009 *Ershi Shiji chu Kang Youwei Baohuanghui zai Meiguo Huaqiao Shehui zhong de Huodong* (Activities of Kang Youwei and the Baohuanghui among the Chinese in the United States in the first part of the 20th century). Xueyuan Chubanshe, Beijing.

Gardner, Albert Dudley

2000 *Two Paths One Destiny: A Comparison of Chinese Households and Communities in Alberta, British Columbia, Montana, and Wyoming, 1848–1910*. Doctoral dissertation, University of New Mexico.

Geertz, Clifford

1961 Studies in Peasant Life: Community and Society. *Biennial Review of Anthropology*, 2:1–41.

Giles, Herbert Allen

1882 *Historic China and Other Sketches*. Harvard University, Harvard CT.

*Glendive Independent*

1887 Citizens of Livingston Order Chinese to Leave. January 29, p. 2.

1894 Revenue Collector Registers Chinese. January 20, p. 2.

Glick, Carl, and Hong Sheng-hwa

1947 Swords of Silence: Chinese Secret Societies Past and Present. *Far Eastern Survey* 16(21):251.

Gluckman, Max

1954 *Rituals of Rebellion in South-East Africa*. Manchester University Press, Manchester, England.

Godelier, Maurice

1986 *The Mental and the Material: Thought, Economy, and Society*. Verso Books, London.

Goldberg, Ray

2006 *Drugs across the Spectrum*. Thompson-Wadsworth Publishing, Belmont CA.

Goldstein, Leslie Friedman

1988 *The Constitutional Rights of Women: Cases in Law and Social Change*. University of Wisconsin, Madison.

Grant, Johnny

1996 *Very Close to Trouble: The Johnny Grant Memoir*. Washington State University Press, Pullman.

Grant, Marilyn

1998 *A Guide to Historic Virginia City*. Montana Historical Society Press, Helena.

Grasso, June M., Jay P. Corrin, and Michael Cort

2004 *Modernization and Revolution in China: From the Opium Wars to World Power*. M. E. Sharpe, New York NY.

*Great Falls Tribune Daily*
1903 Why There Is Not a Chinaman in the City of Great Falls. December 20, p. 10.

Greenwood, Roberta S.
1978 The Overseas Chinese at Home: Life in a Nineteenth-Century Chinatown in California. *Archaeology* 31(4):73–86.
1993 Old Approaches and New Directions: Implications for Future Research. In *Hidden Heritage: Historical Archaeology of the Overseas Chinese*, Priscilla Wegars, ed., pp. 375–403. Baywood Publishing Company, Amityville NY.
1996 *Down by the Station: Los Angeles Chinatown, 1880–1933*. Monumentia Archaeologica, book 18. Cotsen Institute of Archaeology Press, University of California, Los Angeles.

Guotu, Zhuang
2013 China's Policies on Chinese Overseas: Past and Present. In *Routledge Handbook of the Chinese Diaspora*, Tan Chee-Bang, ed., pp. 31–41, Routledge, New York.

Gust, Sherri M.
1993 Animal Bones from Historic Urban Chinese Sites: A Comparison of Sacramento, Woodland, Tucson, Ventura, and Lovelock. In *Hidden Heritage: Historical Archaeology of the Overseas Chinese*, Priscilla Wegars, ed., pp. 117–212. Baywood Publishing Company, Amityville NY.

Gyory, Andrew
1998 *Closing the Gate: Race, Politics, and the Chinese Exclusion Act*. University of North Carolina Press, Chapel Hill.

Hagerman-Benton, Sherry
2007 Remember When? *Sanders County Ledger*. March 15, p. 13.

Hahn, Margie E.
1986 *In Retrospect: A History of Mineral County*. Superior Printing, Superior MT.

Hall, Daniel S., Susan L. Knudsen, Richard D. Gibbons, John Fielding, and Jennifer H. Ogborne
2006 *Marysville Data Recovery: Chinese Laundry Locations, Marysville, Montana, Lewis & Clark County*. Report prepared for Montana Department of Transportation. Western Cultural, Missoula MT.

Hansen, Lawrence Douglas Taylor
2006 The Chinese Six Companies of San Francisco and the Smuggling of Chinese Immigrants across the U.S.-Mexico Border, 1882–1930. *Journal of the Southwest* 48(1):37–61.

Hao Hsing
1893 Letter from Hao Hsing to Ah Hei. Montana Historical Society Manuscript Collection, MC1794, box 1, folder 2. Original in Chinese, translation included in materials. Montana Historical Society, Helena.

Hartman, Howard L.
1992 *SME Mining Engineering Handbook, Volume 1*. Society for Mining Metallurgy and Exploration, Littleton CO.

Hattori, Eugene M., Mary K. Rusco, and Donald R. Tuohy
1979 *Archaeological and Historical Studies at Ninth and Amherst, Lovelock, Nevada*. Archaeological Services Reports. Nevada State Museum, Carson City NV.

Helterline, Maurice
1984 *Horse Plains, Montana Territory*. The Printery, Plains MT.

*Helena Daily Herald*
1867a Revolting Murder. September 10, p. 2.
1867b Strangled. October 24, p. 2.
1867c China Item. December 19, p. 3.
1869 Chinamen for Montana. August 12, p. 3.
1870a Pioneers of Walla Walla. January 28, p. 3.
1870b What's This? September 12, p. 3.
1870c The Number of Chinese . . . October 25, p. 4.
1870d Fire Averted in Missoula. November 11, p. 3.
1871a Chinaman Shot. March 29, p. 3.
1871b Refused Bail. April 10, p. 3.
1871d Ophir Gulch. May 9, p. 3.
1871e Proposed Fire Plan. May 12, p. 3.
1871f The Trial of Martel. May 20, p. 3.
1871g China Case. May 22, p. 3.
1871h The Big Prickly Pear Mines. June 9, p. 3.
1871i Chinamen Found Hung. December 12, p. 4.
1871j Chinese Exclusion. December 27, p. 3.
*Helena Independent*
1875 The Fire Alarm. December 22, p. 3.
1876a Yreka. July 30, p. 3.
1876b Status of German. August 18, p. 3.
1876c Eustace Gulch. September 17, p. 3.
1877a Freight Shipments. May 26, p. 3.
1877b Freight Shipments. June 19, p. 3.
1877c Chinawoman. July 6, p. 3.
1878 Fort Benton Steamer's Manifest. August 8, p. 3.
1880 Quizzing the Pigtails. October 14, p. 3.
1881a Han Bo. March 9, p. 3.
1881b The City Charter Question. June 21, p. 3.
1881c Opium Smoking. June 22, p. 3.
1881d Chinese Marriage. July 15, p. 3.
1881e Celestial Citizens. September 24, p. 3.
1881f John Chinaman . . . September 28, p. 3.
1881g Too Much Opium. October 4, p. 3.
1881h A Chinaman named Oolong. October 23, p. 3.
1881i The Chinese of Butte. December 10, p. 3.
1881j Officer Thom. December 30, p. 3.
1882a Another Heathen. January 11, p. 3.
1882b The Chinaman Who Was Arrested. January 12, p. 3.
1882c Pardoned. January 14, p. 3.
1882d Three More Chinese Opium Smokers. February 4, p. 3.
1882e Chinaman Dies on Steamship. March 30, p. 3.
1882f A Man Named Meagher. July 18, p. 3.
1882g The Chinaman Who Was Captured. July 27, p. 3.
1882h Northern Pacific Construction. August 5, p. 3.
1882i Helena Real Estate. December 2, p. 3.

1882j A Dastardly Murder. December 21, p. 3.
1883a Last Week Three Chinaman. February 14, p. 3.
1883b There Is Scarcely a Town. June 10, p. 3.
1883c Ten Persons, Chinese and Whites. June 13, p. 3.
1883d Next Week's Executions. August 7, p. 3.
1884a Five "Dead Broke" Fellows at Livingston. January 19, p. 3.
1884b 1,000 Chinese Buried. June 7, p. 3.
1884c Chinese Head-Man Dies. June 24, p. 3.
1884d Chin Cum. July 11, p. 3.
1894a The Bones of 28 Chinamen. January 19, p. 3.
1894b San Francisco, Chinese Six Companies. January 24, p. 3.
1894c Up to Present. March 17, p. 3.
1894d Time Is Almost Up. April 11, p. 3.
1894e Deputy Internal Revenue Collector. April 12, p. 3.
1924 Broke but Honest Chinaman Departs. December 22, p. 3.
1925a Yee Sing. April 8, p. 3.
1925b Aged Chinese Leaves Forsyth Home. April 21, p. 3.
1926a Destitute Chinese Released. March 4, p. 3.
1926b Leaves for China. November 5, p. 3.
1927a Hears from China. March 12, p. 3.
1927b An Alder Gulch Pioneer. October 22, p. 3.
1928a Chinese Vegetable Ranchman Murdered. January 4, p. 4.
1928b Clues Point Strongly to Wong Chuck. January 5, p. 3.
1928c New Circumstance Suggests Wong Chuck. January 6, p. 4.
1928d Still Checking Up Chinese. January 7, p. 3.
1928e Wong Chuck, Chinese. January 10, p. 4
1928f Body of Chinese Held in Murder Probe. January 11, p. 4.
1928g Murdered Chinese Given Regulation Funeral. January 13, p. 4.
1928h Wong Chuck Released. January 17, p. 4.
1928i Chinese Murder Seems to Roused Chinese Interest. January 20, p. 4.
1928j R. Woodside Gets His Name Before the Public Again. January 29, p. 4.
1928k Wong Chuck, the Chinese Murder Suspect. February 5, p. 3.
1928l Old Chinaman Dies. February 20, p. 3.
1928m Three Jailed in Narcotic Raids. October 17, p. 4.
1928n Old Time Chinese Expires. November 24, p. 3.
1928o Wong Chuck and Young Ham. December 29, p. 4.
1929a Wong Chuck Pleads Guilty. January 30, p. 4.
1929b Willie Toy, Helena Chinaman. April 1, p. 4.
1929c Aged Chinaman Was Veteran Prospector. April 14, p. 3.
1932 Chinese Population of Philipsburg Dead. July 17, p. 3.
1933a Aged Chinaman Will Be Deported. April 2, p. 3.
1933b Solons Bid Aged Chinese at Choteau Farewell. August 5, p. 3.
1933c Ancient Chinaman Honored at Dinner. August 16, p. 3.
1935 Year Old Body of Chinaman. July 19, p. 3.
1938a Wife of Chinaman at Thompson Falls Killed. January 7, p. 3.
1938b Lone Survivor of Chinese Village at Dillon Passes. January 12, p. 3.

*Helena Weekly Herald*

1867 A China Funeral. September 19, p. 4.

Hilden, Vivian

1942 Chinese Boy Arrives in Montana in Stirring Pioneer Period, Praises Benefactors Opportunities Given by Land of Adoption. *Cut Bank Pioneer Press*, September 25.

Hodder, Ian

1991 Postprocessual Archaeology. In *Processual and Postprocessual Archaeologies: Multiple Ways of Knowing the Past*, Robert W. Preucel ed., pp. 30–41. Southern Illinois University Press, Carbondale.

Hsiao, Kung-Chuan

1960 *Rural China: Imperial Control in the Nineteenth Century*. University of Washington Press, Seattle.

Hsu, Francis

1967 *Under the Ancestor's Shadow: Kinship, Personality, and Social Mobility in China*. Stanford University Press, Stanford CA.

Hsu, Madeline

2003 Unwrapping Orientalist Constraints: Restoring Homosocial Normativity to Chinese American History. *Amerasia Journal* 29(2):230–253.

2006 Trading with Gold Mountain: Jinshanzhuang and Networks of Kinship and Native Place. In *The Flow of People, Resources, and Ideas Between China and America during the Exclusion Era*, Sucheng Chan, ed., pp. 22–33. Temple University Press, Philadelphia PA.

Hunt-Jones, Patricia

2006 *The Heart of a Community: An Archaeological and Historical Study of Island Mountain's Chinese General Store*. Master's of Arts thesis, Department of Anthropology, University of Nevada, Reno.

Ichioka, Yuji

1980 Japanese Immigrant Labor Contractors and the Northern Pacific and the Great Northern Railroad Companies, 1898–1907. *Labor History* 21(3):325–350.

*Idaho Tri-Weekly Statesman*

1882 Rebellious Railroad Hands. October 21, p. 1.

James, Ronald M.

1998 *The Roar and the Silence: A History of Virginia City and the Comstock Lode*. University of Nevada Press, Reno.

James, Ronald M., Richard D. Adkins, and Rachel J. Hartigan

1994 Competition and Coexistence in the Laundry: A View of the Comstock. *Western Historical Quarterly* 25(2):165–184.

Jefferson County Courthouse

1871 Deed Book H. Jefferson County Courthouse, Boulder MT.

Jew, Victor

2002 Exploring New Frontiers in Chinese American History: The Anti-Chinese Riot in Milwaukee, 1889. In *The Chinese in America: A History from Gold Mountain to the New Millennium*, Susie L. Cassel, ed., pp. 77–90. AltaMira Press, Walnut Creek CA.

Jiang, Sun

2004 Imagined Blood: The Creation of Community of Memory through Sworn Brotherhood. *Chinese Sociology and Anthropology* 37(2–3):9–52.

Jung, John

2008 *Chinese Laundries: Tickets to Survival on Gold Mountain.* Privately published.

Kerri, James Nwannukwu

1976 Studying Voluntary Associations as Adaptive Mechanisms: A Review of Anthropological Perspectives. *Current Anthropology* 17(1):23–34.

Kidston, Martin J.

2009 Montana Rail Link to Spend Millions on Mullan Tunnel. *Independent Record*, June, 22, p. 1A.

Kraus-Friedberg, Chana

2008 Transnational Identity and Mortuary Material Culture: The Chinese Plantation Cemetery in Pahala, Hawai'i. *Historical Archaeology* 42(3):123–135.

Kuah-Pearce, Khun-Eng, and Evelyn Hu-Dehart

2006 *Voluntary Organizations in the Chinese Diaspora: Illusions of Open Space in Hong Kong, Tokyo, and Shanghai.* Hong Kong University Press, Hong Kong.

Kuijt, Ian

2001 Place, Death and the Transmission of Social Memory in Early Agricultural Communities of the Near Eastern Pre-Pottery Neolithic. In *Social Memory, Identity, and Death: Anthropological Perspectives on Mortuary Rituals*, pp. 80–99. Archeological Papers of the American Anthropological Association, no. 10.

Lai, David Chuenyan

1991 *The Forbidden City within Victoria.* Orca Press, Victoria, British Columbia.

Lambert, Kirby, Patricia Burnham, and Susan R. Near

2002 *Montana's State Capitol: The People's House.* Montana Historical Society, Helena.

Landreth, Keith, Keo Boreson, and Mary Condon

1985 Archaeological Investigations at the Cabinet Landing Site (10BR413), Bonner County, Idaho. Eastern Washington University Reports in Archaeology and History, Report no. 100–45, Cheney WA.

Lawrence, Susan

2000 *Dolly's Creek: An Archeology of a Victorian Goldfields Community.* Melbourne University Press, Melbourne, Australia.

Lee, Erika

2006 Defying Exclusion: Chinese Immigrants and Their Strategies During the Exclusion Era. In *Chinese American Transnationalism: The Flow of People, Resources, and Ideas Between China and America during the Exclusion Era*, Sucheng Chan, ed., pp. 1–21. Temple University Press, Philadelphia PA.

Lee, Rose Hum

1949 Occupational Invasion, Succession, and Accommodation of the Chinese of Butte, Montana. *American Journal of Sociology* 55(1):50–58.

1960 *The Chinese in the United States of America.* Hong Kong University Press, Hong Kong.

1978 *The Growth and Decline of Chinese Communities in the Mountain Region.* Arno Press, New York NY.

Lee, Ta-Ling

1970 *Foundations of the Chinese Revolution, 1905–1912.* St. John's University Press, New York NY.

Leeson, Michael A.

1885 *History of Montana: 1735–1885.* Warner, Beers and Company, Chicago IL.

Leeson-Vanek, Mona

1996 *Behind These Mountains: Noxon, Heron, Trout Creek, Bull River, and Smeads.* Statesman Press, Spokane WA.

Levi-Strauss, Claude

1963 *Structural Anthropology.* Basic Books, New York NY.

Lewis and Clark County

1866 Ah Gih. Deed Book B, p. 189. Lewis and Clark County Courthouse, Helena MT.

1866 Ah Gown et al., Ah Quay et al., and Ah Rey et al. Deed Book B, p. 192. Lewis and Clark County Courthouse, Helena MT.

Lewty, Peter J.

1987 *To the Columbia Gateway: The Oregon Railway and Northern Pacific, 1879–1884.* Washington State University Press, Pullman.

Li, Wei (ed.)

2006 *From Urban Enclave to Ethnic Suburb: New Asian Communities in Pacific Rim Countries.* University of Hawaii Press, Honolulu.

2009 *Ethnoburb: The New Ethnic Community in Urban America.* University of Hawaii Press, Honolulu.

Liew, K. S.

1985 Some Reflections on Political Change, 1895–1916. In *Ideal and Reality: Social and Political Change in Modern China, 1860–1949,* David Pong and Edmund S.K. Fung, eds., pp. 251–286. University Press of America, Lanham MD.

Lightfoot, Kent

2005 *Indians, Missionaries, and Merchants: The Legacy of Colonial Encounters on the California Frontiers.* University of California Press, Los Angeles.

Lindau, Paul, and Frederic Trautmann

1985 Across Montana on the Northern Pacific in 1883. *Montana: The Magazine of Western History* 35(2):60–65.

Ling, Huping

1988 *Surviving on the Gold Mountain: A History of Chinese American Women and Their Lives.* SUNY Press, Albany NY.

Lister, Florence C., and Robert H. Lister

1989 Chinese Sojourners in Territorial Prescott. *Journal of the Southwest* 31(1):1–111.

Little, Kenneth

1967 Voluntary Associations in Urban Life: A Case Study of Differential Adaptation. In *Social Organization: Essays Presented to Raymond Firth,* pp. 153–167. Aldine Publishing Company, Chicago IL.

Liu, Hong

2005 New Migrants and the Revival of Overseas Chinese Nationalism. *Journal of Contemporary China* 43(14):291–316.

Longenecker, Julia G., and Darby C. Stapp

1993 The Study of Faunal Remains from an Overseas Chinese Mining Camp in Northern Idaho. In *Hidden Heritage: Historical Archaeology of the Overseas Chinese,* Priscilla Wegars, ed., pp. 97–122. Baywood Publishing Company, Amityville NY.

Lydon, Jane

1999 Pidgin English: Historical Archaeology, Cultural Exchange and the Chinese in the Rocks, 1890–1930. In *Historical Archaeology: Back from the Edge,* Pedro Paulo A. Funari, Martin Hall, and Sian Jones, eds., pp. 255–253. Routledge, London.

Lyman, Stanford M.

1974 Conflict and the Web of Group Affiliation in San Francisco's Chinatown, 1850–1910. *Pacific Historical Review* 43(4):473–499.

Lyman, Stanford M., W. E. Willmott, and Berching Ho

1964 Rules of a Chinese Secret Society in British Columbia. *Bulletin of the School of Oriental and African Studies, University of London* 27(3):530–539.

Malone, Michael P.

1989 Beyond the Last Frontier: Toward a New Approach to Western American History. *Western Historical Quarterly* 20(4):409–427.

Malone, Michael P., and Richard Etulain

1989 *The American West*. University of Nebraska Press, Lincoln.

Malone, Michael P., Richard B. Roeder, and William L. Lang

1991 *Montana: A History of Two Centuries*. University of Washington Press, Seattle.

Mann, Susan

1987 *Local Merchants and the Chinese Bureaucracy, 1750–1950*. Stanford University Press, Stanford CA.

Marshall, Douglas A.

2002 Behavior, Belonging, and Belief: A Theory of Ritual Practice. *Sociological Theory* 20(3):360–380.

McCourt, Kathleen A.

2010 VC-21: Green Front Buildings (V094 & V095): A Cultural Resource Survey of Virginia City, Montana (24MA0723). Report prepared for Montana Heritage Preservation and Development Commission, Virginia City.

McGowan, Barry

2005 Chinese Market Gardens in Southern and Western New South Wales. *Australian Humanities Review* 36:1–16.

McGuire, Randall H.

1992 *A Marxist Archaeology*. Percheron Press. Clinton-Corners NY.

McKeown, Adam

2001 *Chinese Migrant Networks and Cultural Change: Peru, Chicago, and Hawai'i, 1900–1936*. University of Chicago Press, Chicago IL.

McLain, Charles

1994 *Chinese Immigrants and American Law*. Routledge, London.

Merritt, Chris

2006 Preliminary Results of Fieldwork: Poacher Gulch Terraces, Fall 2006. Report to Lolo National Forest, Missoula, from Department of Anthropology, University of Montana, Missoula.

2007 The Cedar Creek Chinese: Report on Excavations. Report to Lolo National Forest, Missoula, from Department of Anthropology, University of Montana, Missoula.

2009a The Cedar Creek Chinese: Report on Additional Excavations, Summer of 2008. Report to Lolo National Forest, Missoula, from Department of Anthropology, University of Montana, Missoula.

2009b Final Results of Fieldwork: Poacher Gulch Terraces, Summer 2007. Report to Lolo National Forest, Missoula, from Department of Anthropology, University of Montana, Missoula.

2010 The Chinese on Montana's National Forests: Final Report of Historical and Archaeological Investigations 2006–2010. Report to Regional Office, Region 1, United States Forest Service, Missoula, from Department of Anthropology, University of Montana, Missoula.

Merritt, Christopher W., and Jono L. Mogstad

2009 Final Report on Yogo Town Archaeological Testing. Report to Lewis and Clark National Forest, Great Falls, from University of Montana, Missoula.

Merritt, Christopher W., and C. Milo McLeod

2010 Hand-Stacked Rock Terraces of Sanders County, Montana: The Archaeology and History of Poacher Gulch (24SA0122). *Archaeology in Montana* 51(2).

Merritt, Christopher W., Kassy Marjerrison, and Katie Baumler

2008 Summary of "Funs" Trays Found at China Gulch (24MN0262). *Asian American Comparative Collection Newsletter* 25(4):2.

Merritt, Christopher W., Kelly J. Dixon, and Gary Weisz

2012 "Verily the Road Was Built with Chinaman's Bones": An Archaeology of Chinese Line Camps in Montana. *International Journal of Historical Archaeology* 16(4):666–695.

Meyer, Garren J.

2001 *A Culture History of the German Gulch Chinese*. Government Printing Office, Washington DC.

*Mineral Independent*

1919 Chinese War Savings Society. July 3, p. 4.

1922 Cedar Creek John Dead. March 16, p. 4.

1930 Chinese Cowboy an Old Timer. March 20, p. 4.

1938 Tong War Held in Alder Gulch. February 2, p. 4.

1941 Chinese Waged War at Virginia City. June 12, p. 4.

Mineral Survey Notes

1877 Mineral Plat of Placer Mining Company of Joseph D. Conrad, William Angel, Samuel A. Hanley and Alson E. Ingersoll. Department of the Interior, Bureau of Land Management, Washington, DC. Mineral Survey #547.

*Missoula and Cedar Creek Pioneer*

1870a Chinese Newspaper, September 22, p. 3.

1870b Down on Chinaman, November 10, p. 3.

1870c From Cedar Creek, November 17, p. 2.

1871a John Martelle Trial. May 25, p. 2.

1871b Once More. December 23, p. 2.

1872a The Alien Bill. January 6, p. 2.

1872b Daring Robbery. July 6, p. 2.

1872c From Cedar. July 20, p. 2.

1872d Robbers Captured. July 27, p. 2.

*Montana Daily Record*

1908 Chinamen Are Leaving City. April 13, p. 1.

*Montana Post*

1865 Celestial Delegation. Virginia City, Montana. June 3, p. 2.

*Montana Radiator*

1866 Good Chinamen. February 17, p. 1.

*Montana Standard*

1973 Nevada City Growing as Oldest New Town. June 3, p. 22.

2009 Beal Mountain Mine Reclamation Ongoing. August 14, p. 1.

Morris, Patrick F.

1997 *Anaconda, Montana: Copper Smelting Boom Town on the Western Frontier*. Swann Publishing, Bethesda MD.

Morrow, Trelle A.

2009 Stoneware for Body and Soul: A Social Interpretation of the Chinese Stoneware Record in British Columbia, 1858–1958. Master's thesis, Interdisciplinary Studies, University of British Columbia, Okanagan.

Moschelle, Justin

2009 Master's Project: Big Timber Chinatown Investigations. Professional Portfolio, Department of Anthropology, University of Montana, Missoula.

Moses, Vincent, and Kate Whitmore

1987 Table of the Dragon: Foodways of Overseas Chinese on Riverside. In *Wong Hu Leun: An American Chinatown*, 2 vols., pp. 2:249–258. Great Basin Foundation, San Diego CA.

Mullins, Paul R.

2008 "The strange and unusual": Material and Social Dimensions of Chinese Identity. *Historical Archaeology* 42(3):152–157.

National Park Service

2010 Hardscrabble: The House that Grant Built. http://www.nps.gov/ulsg/historyculture/hardscrabble.htm, accessed 3/21/2010.

*New Northwest*

1869a A Little Row in Deer Lodge. November 26, p. 3.

1869b Big Mines Reported Struck! December 10, p. 3.

1870a A Wedding. April 22, p. 3.

1870b Chinese Getting Blackfoot. June 10, p. 3.

1870c Cedar Creek Chinaman. October 7, p. 3.

1870d Chinese Store. October 28, p. 4.

1871a Fatal Affray at Blackfoot. March 21, p. 3.

1871b $600 Chinawoman. March 21, p. 3.

1871c Negotiations for Sale of Claims in German. April 21, p. 3.

1871d Suicided. May 5, p. 3.

1871e How Is It? June 3, p. 3.

1871f German Gulch. July 15, p. 3.

1871g German Gulch. August 19, p. 3.

1871h Lynching at Henderson. November 4, p. 4.

1872a Gave Up. January 13, p. 3.

1872b The Alien Law, January 27, p. 4.

1872c The Other Side of the Chinese Question. January 27, p. 4.

1872d A Celestial Fireproof. April 6, p. 3.

1872e Blackfoot. June 1, p. 3.

1872f Progress in German. June 15, p. 4.

1872g From Forest City. July 20, p. 4.

1872h Up the Valley and Over the Mountains. August 24, p. 3.

1872i An Episode. August 31, p. 3.

1872j The Heathen Chinee Is Peculiar. September 28, p. 3.
1872k At Horse Prairie . . . October 12, p. 3.
1873a Buy! Buy!! Buy!!! May 17, p. 4.
1873b A Jailable Jubilee. May 31, p. 3.
1873c A Wedding and the Sequel—Two Scenes in a Chinese Drama. June 7, p. 3.
1873d A Horrible Murder on Silver Bow Creek. July 12, p. 3.
1873e No Chinese Wish to Live on Silver Bow Creek. July 26, p. 3.
1873f Sentenced for Life. October 11, p. 3.
1873g Married a Chinawoman. November 22, p. 3.
1874a Disastrous Fire in Helena. January 17, p. 3.
1874b The Chinese. January 17, p. 3.
1874c Yreka Chinawoman. May 9, p. 3.
1874d Dispute, Gem. July 18, p. 3.
1874e Call for Help, Gem. July 25, p. 3.
1874f Letter of Apology. August 8, p. 3.
1874g Condensed Heather. October 17, p. 3.
1874f Ah Kane. November 21, p. 3.
1880a We Have It; An Oriental Vice Enshrined in Deer Lodge. January 16, p. 3.
1880b Raid on Butte Opium Den. March 12, p. 3.
1880c Gem Kee arrives in Deer Lodge. November 5, p. 3.
1880d The Little Foot Wife of Gem Kee. November 28, p. 3.
1880e Buck Gim the Partner of Gem Kee. December 17, p. 3.
1881a The Opium Act. April 15, p. 3.
1881b Quong Lee. August 5, p. 3.
1881c John Dunn Found Dead. October 7, p. 3.
1882a Butte and Vicinity. May 12, p. 3.
1882b Hallett Had 4,000 Chinamen. August 11, p. 3.
1882c The Western Invasion. December 8, p. 3.
1883 On the Road. November 2, p. 3.
1884 A Raid. December 5, p. 3.
1885a Chinese Gardener of Peterson Creek. March 27, p. 3.
1885b Coroner McFarland. April 17, p. 3.
1885c Giant Powder at Anaconda. April 17, p. 3.
1885d If "the Chinese must go." September 18, p. 3.
1885e The Butte Papers. September 25, p. 3.
1885f The Chinese Matter. October 2, p. 3.
1885g Another Precinct Heard From. October 16, p. 3.
1885h Boycotting. October 30, p. 3.
1885i Boycotted. December 4, p. 3.

*New York Times*
1882 Oregon Men and Politics. July 7, p. 3.
1883 Ten Men Lynched. February 6, p. 3.

Newby, Rick
1987 Helena's Social Supremacy: Political Sarcasm and the Capital Fight. *Montana: The Magazine of Western History* 37(4):68–72.

Nolan, Edward W.

1983 "Not without Labor and Expense": The Villard–Northern Pacific Last Spike Excursion, 1883. *Montana: The Magazine of Western History* 33(3):2–11.

Norman, William

2012 Tradewinds and Traditions: Exploring the Archaeology of German Gulch. Master's thesis, Department of Anthropology, University of Montana, Missoula.

Olsen, John W.

1978 A Study of Chinese Ceramics Excavated in Tucson. *Kiva* 44(2):1–50.

1983 An Analysis of East Asian Coins Excavated in Tucson, Arizona. *Historical Archaeology* 17(2):41–55.

O'Neill Richard, Amy

1999 *International Trafficking in Women to the United States: A Contemporary Manifestation of Slavery and Organized Crime*. Center for the Study of Intelligence, Department of State, Washington DC.

Ownby, David

1996 *Brotherhoods and Secret Societies in Early and Mid-Qing China*. Stanford University Press, Stanford CA.

2001 Recent Chinese Scholarship on the History of Chinese Secret Societies. *Late Imperial China* 22(1):139–158.

Ownby, David, and Mary Somers Heidhues

1993 *"Secret Societies" Reconsidered: Perspectives on the Social History of Modern South China and Southeast Asia*. M. E. Sharpe, New York NY.

Paciotti, Brian

2005 Homicide in Seattle's Chinatown, 1900–1940: Evaluating the Influence of Social Organizations. *Homicide Studies* 9(3):229–255.

Paige, Ronald B., Dennis C. Darling, and Jack B. McCoy

1998 *Ghost Town Trails: A Guide to the Historic Ghost Towns of Granite County, Montana*. Ronald B. Paige, Philipsburg MT.

Pan, Lynn

1994 *Sons of the Yellow Emperor: A History of the Chinese Diaspora*. Kodansha International, New York NY.

Parliament of the Commonwealth of Australia (PCA)

1995 *Asian Organised Crime in Australia*. A Discussion Paper by the Parliamentary Joint Committee on the National Crime Authority, Commonwealth of Australia, Sydney.

Pegler-Gordon, Anna

2006 Chinese Exclusion, Photography, and the Development of the U.S. Immigration Policy. *American Quarterly* 58(1):51–77.

Pfaelzer, Jean

2007 *Driven Out: The Forgotten War against Chinese Americans*. Random House, New York NY.

Phillips, Paul C. (ed.)

1925 *Forty Years on the Frontier as Seen in the Journals and Reminisces of Granville Stuart*. Arthur H. Clark Company, Cleveland OH.

Ping-ti, Ho

1959 *Studies on the Population of China, 1368–1953*. Harvard University Press, Cambridge MA.

1962 *The Ladder of Success in Imperial China: Aspects of Social Mobility, 1368–1911*. Columbia University Press, New York NY.

*Pittsburgh Press*

1903 Tommy Haw's Check Is Good. February 20, p. 3.

Praetzellis, Adrian, and Mary Praetzellis

2001 Mangling Symbols of Gentility in the Wild West. *American Anthropologist* 103(3):645–654.

Radcliffe-Brown, A. R.

1952 *Structure and Function in Primitive Society: Essays and Addresses*. Routledge, London.

Raymond, Rossiter W.

1869 *The Mines of the West: A Report to the Secretary of the Treasury*. J. B. Ford and Company, New York NY.

1870 *Statistics of Mines and Mining in the States and Territories West of the Rocky Mountains*. Government Printing Office, Washington DC.

1873 *Silver and Gold: An Account of the Mining and Metallurgical Industry of the United States*. J. B. Ford and Company, New York NY.

*Record-Union*

1892 No Love for the Heathen. January 29, p. 1.

Renz, Louis Tuck

1980 *The History of the Northern Pacific Railroad*. Ye Galleon Press, Fairfield WA.

Rice, David G.

1977 *An Archaeological Assessment of Historic Moose City, Clearwater County, Idaho*. Anthropological Research Manuscript Series, no. 30. University of Idaho, Moscow.

Ritchie, Neville

1986 *Archaeology and History of the Chinese in Southern New Zealand during the Nineteenth Century: A Study of Acculturation, Adaptation, and Change*. Master's thesis, University of Otago, New Zealand.

1993 Form and Adaptation: Nineteenth Century Chinese Miners' Dwellings in Southern New Zealand. In *Hidden Heritage: Historical Archaeology of the Overseas Chinese*, Priscilla Wegars, ed., pp. 335–373. Baywood Publishing Company, Amityville NY.

2003 Taking Stock: 20 years of Australasian "Overseas Chinese Archaeology." *Australasian Historical Archaeology* 21:4–10.

*River Press*

1885 The Chinese Must Go. December 23, p. 2.

Robison, Ken

2009a "Aliens Alive and Aliens Dead": The Celestial Kingdom on the Upper Missouri Part II. *Fort Benton River Press*, November 9, pp. 2–3.

2009b The Celestial Kingdom on the Upper Missouri: The Chinese of Historic Chouteau County—Part I. *Fort Benton River Press*, October 21, p. 1.

2010 The Benton Chinese Persevered and Worked On: The Celestial Kingdom on the Upper Missouri—Part III. *Fort Benton River Press*, February 24, p. 3.

Roenke, Karl

2000 Addendum to Site File #24MO110 (Montreal town site or Old Town). Ninemile Fire Burned Area Emergency Rehabilitation (*BAER*) observations. On file, Lolo National Forest Supervisor's Office, Missoula MT.

Rohe, Randall E.

1982 After the Gold Rush: Chinese Mining in the Far West. *Montana: The Magazine of Western History* 32(4):2–19.

1996 Chinese River Mining in the West. *Montana: The Magazine of Western History* 46(3):14–29.

Ross, Peter V.

1909 *Probate Law and Practice: A Treatise on Wills, Succession, Administration and Guardianship.* Bancroft-Whitney Company, San Francisco CA.

Rossillon, Mitzi

1997 *Granite: A Cultural Resource Inventory and Evaluation Report.* Renewable Technologies, Butte MT.

2002 *Cave Gulch Wildfire Post-Burn Heritage Resources Assessment, Lewis & Clark and Broadwater Counties, Montana*, Vol. 1. Renewable Technologies, Butte MT.

2003 *Cave Gulch Wildfire Lewis & Clark and Broadwater Counties, Montana: Archaeological Testing and Detailed Recordation*, Vol. 1. Renewable Technologies, Butte MT.

2008 *Final Report of Archaeological Investigations of Butte's Chinatown.* Renewable Technologies, Butte MT.

Rouse, Wendy L.

2005 "What we didn't understand": A History of Chinese Death Ritual in China and California. In *Chinese American Death Rituals: Respecting the Ancestors*, pp. 19–45. AltaMira Press, Lanham MD.

Rusco, Elmer R.

2002 Riot in Unionville, Nevada: A Turning Point. In *The Chinese in America: A History from Gold Mountain to the New Millennium*, Susie L. Cassel, ed., pp.91–105. AltaMira Press, Walnut Creek CA.

Sadowski-Smith, Claudia

2008 Unskilled Labor Migration and the Illegality Spiral: Chinese, European, and Mexican Indocumentados in the United States, 1882–2007. *American Quarterly* 60(3):779–804.

Said, Edward W.

1978 *Orientalism.* Vintage Books, London.

*Salt Lake Herald*

1892 Chinese in Montana. January 28, p. 1.

1893 Crusade against Chinese. February 11, p. 1.

Sando, Ruth Ann, and David L. Felton

1993 Inventory Records of Ceramics and Opium from a Nineteenth Century Chinese Store in California. In *Hidden Heritage: Historical Archaeology of the Overseas Chinese*, Priscilla Wegars, ed., pp. 151–176. Baywood Publishing Company, Amityville NY.

Saxton, Alexander

1975 *The Indispensable Enemy: Labor and the Anti-Chinese Movement in California.* University of California Press, Berkeley.

Schlegel, Gustave

1866 *Thian Ti Hwui, the Hung-League or Heaven-Earth-League: A Secret Society with the Chinese in China and India.* Lange and Company, Batavia.

Schneider, Carrie

2004 Remembering Butte's Chinatown. *Montana: The Magazine of Western History* 54(2):67–69.

Schwantes, Carlos A.
1982 Protest in a Promised Land: Unemployment, Disinheritance, and the Origin of Labor Militancy in the Pacific Northwest, 1885–1886. *Western Historical Quarterly* 13(4):373–390.
Sears, Roebuck & Company
1897 *Sears, Roebuck and Company Catalogue*. Skyhorse Publishing, New York NY.
Shaoqing, Cai
2002 On the Overseas Chinese Secret Societies of Australia. *New Zealand Journal of Asian Studies* 4(1):30–45.
Sheppard, Juniata L.
1909 *Laundry Work for Use in Homes and Schools*. Webb Publishing Company, St. Paul MN.
Smalley, Eugene V.
1883 *History of the Northern Pacific Railroad*. G. P. Putnam and Sons, New York NY.
Smith, Carrie E., and Kelly J. Dixon
2005 Determination of Eligibility for Inclusion in the National Register of Historic Places of 19 Historic Sites within the Heavenly Ski Resort, Douglas County, Nevada. Lake Tahoe Basin Management Unit Report R2004–0519–00048. Unpublished report prepared for U.S. Forest Service, Lake Tahoe Basin Management Unit, South Lake Tahoe and Heavenly Ski Resort, South Lake Tahoe CA.
Smith, Lindsay M.
2003 Identifying Chinese Ethnicity through Material Culture: Archaeological Excavations at Kiandra, NSW. *Australasian Historical Archaeology* 21:18–29.
Smith, Richard J.
1994 *China's Cultural Heritage: The Qing Dynasty, 1644–1912*. Westview Press, Boulder CO.
Smits, Nicholas J.
2008 Roots Entwined: Archaeology of an Urban Chinese American Cemetery. *Historical Archaeology* 42(3):111–122.
Speers, William
1870 *The Oldest and Newest Empire: China and the United States*. S. S. Scranton and Company, Philadelphia PA.
Spence, Clark C.
1975 *Territorial Politics and Government in Montana: 1864–1889*. University of Illinois Press, Urbana.
Spence, Jonathan D.
1990 *The Search for Modern China*. W. W. Norton Company, New York NY.
Spier, Robert F. G.
1958 Tool Acculturation among 19th-Century California Chinese. *Ethnohistory* 5(2):97–117.
*Spokane Falls Review*
1883 Death's Carnival. June 30, p. 1.
Stanin, Zvonkica
2004 From Li Chun to Yong Kit: A Market Garden on the Loddon, 1851–1912. *Journal of Australian Colonial History* 6:15–34.
Stanley, Timothy J.
1996 "Chinamen, Wherever We Go": Chinese Nationalism and Guangdong Merchants in British Columbia, 1871–1911. *Canadian Historical Review* 77(4):475–503.

State of Montana

1909 *Laws, Resolutions and Memorials of the State of Montana Passed at the Eleventh Regular Session of the Legislative Assembly*. Independent Publishing Company, Helena MT.

1953 *Laws, Resolutions and Memorials of the State of Montana Passed at the Thirty-Third Legislative Assembly in Regular Session*. State Publishing Company, Helena MT.

Steward, Julian

1955 *Theory of Culture Change: The Methodology of Multilinear Evolution*. University of Illinois Press, Urbana.

Stirling, W. G., and J.S.M. Ward

2003 [1925] *Hung Society or the Society of Heaven and Earth*. Reprint, Kessinger Publishing, Whitefish MT.

Stoner, John

2007 Broadwater County Chinese: The Forgotten Immigrant 1865–1925. Unpublished Report. On file, Montana Historical Society Archives, Helena.

Sung, Betty Lee

1972 *The Chinese in America*. Macmillan, New York NY.

Swartout, Robert R., Jr.

1988 Kwangtung to Big Sky: The Chinese in Montana, 1864–1900. *Montana: The Magazine of Western History* 38(1):42–53.

2001 Kwangtung to Big Sky: The Chinese in Montana, 1864–1900. In *Chinese on the American Frontier*, Arif Dirlick and Malcolm Yeung, eds., pp. 367–382. Rowman and Littlefield, Lanham MD.

Taylor, Bill, and Jan Taylor

1999 *The Northern Pacific's Rails to Gold and Silver: Lines to Montana's Mining Camps, Vol. 1: 1883–1887*. Pictorial Histories, Missoula MT.

2002 *Northern Pacific's Mullan Pass on the "Montana Short Line."* Pictorial Histories, Missoula MT.

Taylor, Walter W.

1948 *A Study of Archeology*. Southern Illinois University Press, Carbondale.

Teague, George A., and Lynette O. Shenk

1977 *Excavations at Harmony Borax Works* Western Archeological Center Publications in Anthropology, no. 6. National Park Service, Washington DC.

Thompson, Francis M., and Kenneth N. Owens

2004 *A Tenderfoot in Montana: Reminiscences of the Gold Rush, the Vigilantes, and the Birth of Montana Territory*. Globe Peqout Press, Guilford CT.

Toole, K. Ross

1959 *Montana: An Uncommon Land*. University of Oklahoma Press, Norman.

Tsai, Shih-Shan Henry

1986 *The Chinese Experience in America*. Indiana University Press, Bloomington.

U.S. Department of State

1902 *Papers Relating to the Foreign Relations of the United States*. Department of State, Government Printing Office, Washington DC.

United States Army

1881 Register of Burials at Fort Shaw. Records of the United States Army, Continental Commands, Records Group 393.7, accessed via Ancestry.com, May 1, 2008.

*United States v. Foo Duck*

1909 Decisions Involving International Law, U.S. v. Foo Duck. *American Journal of International Law* 3(2):489–491.

Valentine, David

2002 Chinese Placer Mining in the United States: An Example from American Canyon Nevada. In *The Chinese in America: A History from Gold Mountain to the New Millennium*, Susie L. Cassel, ed., pp. 37–53. AltaMira Press, Walnut Creek CA.

Van der Veen, Marijke

2003 When Is Food a Luxury? *World Archaeology* 34(3):405–427.

Vollmann, William T.

2004 They Came Out Like Ants: Searching for the Chinese Tunnels of Mexicali. *Harper's Magazine*, October, pp. 10–30.

Voss, Barbara L.

2005 The Archaeology of the Overseas Chinese Communities. *World Archaeology* 37(3):424–439.

2008 Between the Household and the World System: Social Collectivity and Community Agency in Overseas Chinese Archaeology. *Historical Archaeology* 42(3):37–52.

Voss, Barbara L., and Rebecca Allen

2008 Overseas Chinese Archaeology: Historical Foundations, Current Reflections, and New Directions. *Historical Archaeology* 42(3):5–28.

Voynick, Stephen M.

1985 *Yogo: The Great American Sapphire.* Mountain Press Publishing Company, Missoula MT.

Wakeman, Frederic

1972 The Secret Societies of Kwangtung, 1800–1856. In *Popular Movements and Secret Societies in China 1840–1950*, J. Chesneaux, ed., pp. 29–48. Stanford University Press, Stanford CA.

Warner, W. Lloyd

1963 A Methodology for the Study of Social Class. In *Social Structure: Studies Presented to A. R. Radcliffe-Brown*, ed. Meyer Fortes, pp. 1–17, Russell and Russell, New York.

Warren, Charles S.

1896 Centennial Address on Montana and Butte, delivered July 4, 1876. In *Contributions to the Historical Society of Montana*, Vol. 2. State Publishing Company, Helena MT.

*Weekly Missoulian*

1882a Raid on Opium Den. April 7, p. 3.

1882b Cabinet Landing. May 19, p. 1.

Weisz, Gary J.

2003 *Stepping Light: Revisiting the Construction Camps on the Lake Pend d'Oreille and Clark Fork Division of the Northern Pacific Railroad, 1879–1883, Volume III.* Unpublished Manuscript, on file, Idaho State Historic Preservation Office, Boise.

Wegars, Priscilla

1991 Who's Been Workin' on the Railroad?: An Examination of the Construction, Distribution, and Ethnic Origins of Domed Rock Ovens on Railroad-Related Sites. *Historical Archaeology* 25(1):37–65.

Wegars, Priscilla (ed.)

1993 *Hidden Heritage: Historical Archaeology of the Overseas Chinese.* Baywood Publishing Company, Amityville NY.

*Whitefish Pilot*
1923 Find Opium Den. November 20, p. 3.
1930 Old-Time Whiskey Runner. January 24, p. 3.
Williams, Bryn
2008 Chinese Masculinities and Material Culture. *Historical Archaeology* 42(3):53–67.
Williams, Michael
1999 *Chinese Settlement in New South Wales: A Thematic History*. New South Wales Heritage Office, Sydney.
Williamson, Samuel H.
2010 Seven Ways to Compute the Relative Value of a U.S. Dollar Amount, 1790 to Present. Measuring Worth, http://www.measuringworth.com/uscompare/.
Wolle, Muriel Sibell
1963 *Montana Pay Dirt: A Guide to the Mining Camps of the Treasure State*. Ohio University Press, Athens.
Wong, Roy Bin
1997 *China Transformed: Historical Change and the Limits of European Experience*. Cornell University Press, Ithaca NY.
Woon, Yuen-fong
1984 *Social Organization in South China, 1911–1949: The Case of the Kuan Lineage in K'ai-p'ing County*. Center for Chinese Studies, University of Michigan, Ann Arbor.
Works Progress Administration (WPA)
[1930s] Mining Oral Histories, Superior, MT. Items 72, 75, 86. On file, Mineral County Historical Society, Superior MT.
Wrobleski, David E.
1996 *The Archaeology of Chinese Work Camps on the Virginia and Truckee Railroad*. Master's thesis, Department of Anthropology, University of Nevada, Reno.
Wunder, John R.
1980 Law and Chinese in Frontier Montana. *Montana: The Magazine of Western History*, 30(3):18–31.
Wylie, Jerry, and Richard Fike
1993 Opium Smoking Paraphernalia and Techniques. In *Hidden Heritage: Historical Archaeology of the Overseas Chinese*, Priscilla Wegars, ed., pp. 111–149. Baywood Publishing Company, Amityville NY.
Wynne, Robert E.
1978 *Reaction to the Chinese in the Pacific Northwest and British Columbia: 1855–1910*. Arno Press, New York NY.
Xia, Jingfeng
2001 *Foodways and Their Significance to Ethnic Integration: An Ethnoarchaeological and Historical Archaeology Survey of the Chinese in Tucson, Arizona*. Doctoral dissertation, Department of Anthropology, University of Arizona, Tucson.
Yang, Jeannie K., and Virginia R. Hellmann
1996 What's in the Pot? An Emic Study of Chinese Brown Glazed Stoneware. *Proceedings of the Society for California Archaeology* 11:59–66.
Zelin, Madeleine
1990 The Rise and Fall of the Fu-Rong Salt-Yard Elite: Merchant Dominance in Late Qing China. In *Chinese Local Elites and Patterns of Dominance*, J. W. Esherick and M. B. Rankin, eds., pp. 82–111. University of California Press, Berkeley.

Zhao, Kang

1986 *Man and Land in Chinese History: An Economic Analysis*. Stanford University Press, Stanford CA.

Zheng, Zhenman

2001 *Family Lineage Organization and Social Change in Ming and Qing Fujian*. University of Hawaii Press, Honolulu.

Zhu, Liping

1995 "A Chinaman's Chance" on the Rocky Mountain Mining Frontier. *Montana: The Magazine of Western History* 45(4):36–51.

1997 *A Chinaman's Chance: The Chinese on the Rocky Mounting Mining Frontier*. University Press of Colorado, Boulder.

# Index

*Page numbers in italic indicate illustrations.*

In the Historical Archaeology of the American West series

*Mining Archaeology in the American West: A View from the Silver State*
Donald L. Hardesty

*Eldorado!: The Archaeology of Gold Mining in the Far West*
Edited by Catherine Holder Spude, Robin O. Mills, Karl Gurcke, Roderick Sprague

*On the Edge of Purgatory: An Archaeology of Place in Hispanic Colorado*
Bonnie J. Clark

*Virginia City: Secrets of a Western Past*
Ronald M. James

*Historical Archaeology Through a Western Lens*
Edited and with an introduction by Mark Warner and Margaret Purser

*The Coming Man from Canton: Chinese Experience in Montana, 1862–1943*
Christopher W. Merritt

To order or obtain more information on these or other University of Nebraska Press titles, visit nebraskapress.unl.edu.

CPSIA information can be obtained
at www.ICGtesting.com
Printed in the USA
LVHW101156081222
734810LV00018B/502/J

9 780803 299788